PUBLIC HEALTH IN THE 21ST CENTURY

CELL PHONE USE AND HEALTH RISKS

ASSESSMENTS AND STATE OF RESEARCH

PUBLIC HEALTH IN THE 21ST CENTURY

Additional books in this series can be found on Nova's website under the Series tab.

Additional e-books in this series can be found on Nova's website under the e-books tab.

MEDIA AND COMMUNICATIONS - TECHNOLOGIES, POLICIES AND CHALLENGES

Additional books in this series can be found on Nova's website under the Series tab.

Additional e-books in this series can be found on Nova's website under the e-books tab.

PUBLIC HEALTH IN THE 21ST CENTURY

CELL PHONE USE AND HEALTH RISKS

ASSESSMENTS AND STATE OF RESEARCH

CYNTHIA KEMPSON
AND
EUGENE RAHM
EDITORS

nova publishers
New York

For permission to use material from this book please contact us:
Telephone 631-231-7269; Fax 631-231-8175
Web Site: http://www.novapublishers.com

Additional color graphics may be available in the e-book version of this book.

Library of Congress Cataloging-in-Publication Data

ISBN: 978-1-62257-948-8

Published by Nova Science Publishers, Inc. † New York

CONTENTS

PREFACE

Mobile phone use in the United States has risen dramatically over the last 20 years, and Americans increasingly rely on mobile phones as their sole or primary means of telephone communication. The rapid adoption of mobile phones has occurred amidst controversy over whether the technology poses a risk to human health. Like other devices that transmit radio signals, mobile phones emit radio-frequency (RF) energy. At high power levels, RF energy can heat biological tissue and cause damage. Though mobile phones operate at power levels well below the level at which this thermal effect occurs, the question of whether long-term exposure to RF energy emitted from mobile phones can cause other types of adverse health effects, such as cancer, has been the subject of research and debate. This book examines what is known about the health effects of RF energy from mobile phones, with a focus on the FCC and FDA's regulatory responsibilities; and other scientific research.

Chapter 1 – The rapid adoption of mobile phones has occurred amidst controversy over whether the technology poses a risk to human health as a result of long-term exposure to RF energy from mobile phone use. FCC and FDA share regulatory responsibilities for mobile phones. GAO was asked to examine several issues related to mobile phone health effects and regulation. Specifically, this report addresses (1) what is known about the health effects of RF energy from mobile phones and what are current research activities, (2) how FCC set the RF energy exposure limit for mobile phones, and (3) federal agency and industry actions to inform the public about health issues related to mobile phones, among other things. GAO reviewed scientific research; interviewed experts in fields such as public health and engineering, officials from federal agencies, and representatives of academic institutions, consumer groups, and the mobile phone industry; reviewed mobile phone testing and certification regulations and guidance; and reviewed relevant federal agency websites and mobile phone user manuals.

Chapter 2 – *Key Points:*

- Cell phones emit radiofrequency energy, a form of non-ionizing electromagnetic radiation, which can be absorbed by tissues closest to where the phone is held.
- The amount of radiofrequency energy a cell phone user is exposed to depends on the technology of the phone, the distance between the phone's antenna and the user, the extent and type of use, and the user's distance from cell phone towers.

Studies thus far have not shown a consistent link between cell phone use and cancers of the brain, nerves, or other tissues of the head or neck. More research is needed because cell phone technology and how people use cell phones have been changing rapidly.

Chapter 3 – Statement of Dariusz Leszczynski, Ph.D., Research Professor, Radiation and Nuclear Safety Authority, University of Helsinki.

Chapter 4 – Testimony of Siegal Sadetzki, M.D., Director, Cancer and Radiation Epidemiology Unit, Gertner Institute, Israel.

Chapter 5 – Statement of Linda S. Erdreich, Ph.D, Senior Managing Scientist, Exponent Health Sciences.

Chapter 6 – Statement of Devra Lee Davis, Ph.D., Professor of Epidemiology, University of Pittsburgh.

Chapter 7 – Testimony of Olga V. Naidenko, Ph.D., Senior Scientist, Environmental Working Group.

In: Cell Phone Use and Health Risks
Editors: C. Kempson and E. Rahm
ISBN: 978-1-62257-948-8

Chapter 1

TELECOMMUNICATIONS: EXPOSURE AND TESTING REQUIREMENTS FOR MOBILE PHONES SHOULD BE REASSESSED*

United States Government Accountability Office

WHY GAO DID THIS STUDY

The rapid adoption of mobile phones has occurred amidst controversy over whether the technology poses a risk to human health as a result of long-term exposure to RF energy from mobile phone use. FCC and FDA share regulatory responsibilities for mobile phones. GAO was asked to examine several issues related to mobile phone health effects and regulation. Specifically, this report addresses (1) what is known about the health effects of RF energy from mobile phones and what are current research activities, (2) how FCC set the RF energy exposure limit for mobile phones, and (3) federal agency and industry actions to inform the public about health issues related to mobile phones, among other things. GAO reviewed scientific research; interviewed experts in fields such as public health and engineering, officials from federal agencies, and representatives of academic institutions, consumer groups, and the mobile phone industry; reviewed mobile phone testing and

* This is an edited, reformatted and augmented version of United States Government Accountability Office, Publication No. GAO-12-771, dated July 2012.

certification regulations and guidance; and reviewed relevant federal agency websites and mobile phone user manuals.

What GAO Recommends

FCC should formally reassess and, if appropriate, change its current RF energy exposure limit and mobile phone testing requirements related to likely usage configurations, particularly when phones are held against the body. FCC noted that a draft document currently under consideration by FCC has the potential to address GAO's recommendations.

What GAO Found

Scientific research to date has not demonstrated adverse human health effects of exposure to radio-frequency (RF) energy from mobile phone use, but research is ongoing that may increase understanding of any possible effects. In addition, officials from the Food and Drug Administration (FDA) and the National Institutes of Health (NIH) as well as experts GAO interviewed have reached similar conclusions about the scientific research. Ongoing research examining the health effects of RF energy exposure is funded and supported by federal agencies, international organizations, and the mobile phone industry. NIH is the only federal agency GAO interviewed directly funding studies in this area, but other agencies support research under way by collaborating with NIH or other organizations to conduct studies and identify areas for additional research.

The Federal Communications Commission's (FCC) RF energy exposure limit may not reflect the latest research, and testing requirements may not identify maximum exposure in all possible usage conditions. FCC set an RF energy exposure limit for mobile phones in 1996, based on recommendations from federal health and safety agencies and international organizations. These international organizations have updated their exposure limit recommendation in recent years, based on new research, and this new limit has been widely adopted by other countries, including countries in the European Union. This new recommended limit could allow for more RF energy exposure, but actual exposure depends on a number of factors including how the phone is held during use. FCC has not adopted the new recommended limit. The Office of

Management and Budget's instructions to federal agencies require the adoption of consensus standards when possible. FCC told GAO that it relies on the guidance of federal health and safety agencies when determining the RF energy exposure limit, and to date, none of these agencies have advised FCC to change the limit. However, FCC has not formally asked these agencies for a reassessment. By not formally reassessing its current limit, FCC cannot ensure it is using a limit that reflects the latest research on RF energy exposure. FCC has also not reassessed its testing requirements to ensure that they identify the maximum RF energy exposure a user could experience. Some consumers may use mobile phones against the body, which FCC does not currently test, and could result in RF energy exposure higher than the FCC limit.

Federal agencies and the mobile phone industry provide information on the health effects of mobile phone use and related issues to the public through their websites and mobile phone manuals. The types of information provided via federal agencies' websites on mobile phone health effects and related issues vary, in part because of the agencies' different missions, although agencies provide a broadly consistent message. Members of the mobile phone industry voluntarily provide information on their websites and in mobile-phone user manuals. There are no federal requirements that manufacturers provide information to consumers about the health effects of mobile phone use.

ABBREVIATIONS

ANSI	American National Standards Institute
CDC	Centers for Disease Control and Prevention
DNA	deoxyribonucleic acid
EPA	Environmental Protection Agency
FCC	Federal Communications Commission
FDA	Food and Drug Administration
IARC	International Agency for Research on Cancer
IEEE	Institute of Electrical and Electronics Engineers
NIH	National Institutes of Health
OSHA	Occupational Safety and Health Administration
RF	radio frequency
SAR	specific absorption rate
TCB	Telecommunication Certification Body

The Honorable
Henry A. Waxman
Ranking Member
Committee on Energy and Commerce
House of Representatives

The Honorable
Anna G. Eshoo
Ranking Member
Subcommittee on Communications and Technology
Committee on Energy and Commerce
House of Representatives

The Honorable
Edward J. Markey
House of Representatives

Mobile phone use in the United States has risen dramatically over the last 20 years, and Americans increasingly rely on mobile phones as their sole or primary means of telephone communication.[1] The rapid adoption of mobile phones has occurred amidst controversy over whether the technology poses a risk to human health. Like other devices that transmit radio signals, mobile phones emit radio-frequency (RF) energy. At high power levels, RF energy can heat biological tissue and cause damage. Though mobile phones operate at power levels well below the level at which this thermal effect occurs, the question of whether long-term exposure to RF energy emitted from mobile phones can cause other types of adverse health effects, such as cancer, has been the subject of research and debate.

The Federal Communications Commission (FCC) and the Food and Drug Administration (FDA) share regulatory responsibilities for mobile phones. FCC, in compliance with the National Environmental Policy Act of 1969, regulates RF energy emitted from FCC-regulated transmitters, including mobile phones. 2 Toward that end, FCC has implemented a certification program to ensure that all mobile phones sold in the United States comply with the agency's limit on RF energy exposure. This limit was designed to protect users from the thermal effects of acute exposure to RF energy. FDA is responsible for carrying out a program designed to protect public health and safety from electronic product radiation.[3] FDA does not review the safety of all radiation-emitting electronic products, such as mobile phones, before they are marketed. However, FDA has the authority to take action, such as requiring manufacturers to replace or recall mobile phones that are shown to emit RF energy at a level that is hazardous.[4] To date, FDA has not taken such action, but the agency regularly evaluates scientific studies on mobile phones and health to determine whether they raise public health questions.

In 2001, we reported on the status of scientific knowledge about potential health risks of RF energy exposure from mobile phones and the federal government's regulatory actions to ensure mobile phone safety. found that FDA and others had concluded that the research did not show RF energy exposure from mobile phones had adverse health effects, but more studies were needed. We also found that FCC had not issued standardized procedures for testing mobile phones and that FCC's and FDA's consumer materials could be improved. Since 2001, FCC has issued revised guidance for mobile phone testing, and both FCC and FDA have provided updated information to consumers about the health effects of mobile phone use.

At your request, we are updating information related to mobile phone health effects and regulatory issues. Specifically, this report addresses:

1) What is known about the human health effects of RF energy exposure from mobile phone use, and what are the current research activities of federal agencies and other organizations?
2) How has FCC set the RF energy exposure limit for mobile phones and how does FCC ensure compliance with the limit?
3) What actions have federal agencies and the mobile phone industry taken to inform the public about issues related to mobile phone health effects?

To determine what is known about the human health effects of RF energy exposure from mobile phone use, we reviewed selected studies, including studies and reports that review and assess the scientific research as well as key individual studies. We identified these studies through literature searches in online databases, such as Embase and Medline, and interviews with officials from federal agencies, academic institutions, consumer groups, and industry associations. We also interviewed subject matter experts in a range of fields, such as public health and engineering. To determine the current research activities of federal agencies related to mobile phone use and health, we interviewed officials from FCC; the Department of Health and Human Services' FDA, Centers for Disease Control and Prevention (CDC), and National Institutes of Health (NIH); Department of Defense; Department of

Labor's Occupational Safety and Health Administration (OSHA); and the Environmental Protection Agency (EPA). To determine the research activities of other organizations, we interviewed representatives of the International Agency for Research on Cancer (IARC), academic institutions, consumer groups, mobile phone industry associations, mobile phone manufacturers, and mobile phone providers. To determine how FCC set the RF energy exposure limit and ensures compliance with it, we reviewed FCC regulations and guidance. We also reviewed reports from international organizations that recommend RF energy exposure limits. We conducted interviews with officials from FCC and Telecommunication Certification Bodies (TCBs) to understand their role in certifying mobile phones. We also interviewed representatives of the mobile phone industry and consumer organizations, and experts in RF energy exposure limits to obtain their perspectives on the testing and certification of mobile phones. To determine the actions federal agencies and the mobile phone industry have taken to inform the public about issues related to mobile phone health effects, we reviewed information on the public websites of CDC, EPA, FCC, FDA, NIH, and OSHA. We also reviewed the user manuals for selected top-selling mobile phones of 2011 to identify the information manufacturers provided to consumers. (See app. I for more information on our scope and methodology and app. II for a list of studies we reviewed.)

We conducted this performance audit from August 2011 through July 2012 in accordance with generally accepted government auditing standards. Those standards require that we plan and perform the audit to obtain sufficient, appropriate evidence to provide reasonable basis for our findings and conclusions based on our audit objectives. We believe that the evidence obtained provides a reasonable basis for our findings and conclusions based on our audit objectives.

BACKGROUND

The United States has experienced dramatic changes in mobile phone use since nationwide cellular service became available in the mid-1980s. For example, the number of estimated mobile phone subscribers has grown from about 3.5 million in 1989 to approximately 286 million by the end of 2009, according to the most recent data reported by FCC.[6] Further, the number of Americans who rely exclusively on mobile phones for voice service has increased in recent years. For example, by the end of 2009 over 50 percent of young adults aged 25 to 29 relied exclusively on mobile phones, according to the most recent FCC data.[7] The way individuals use mobile phones has also changed. For instance, while average minutes of use per mobile phone subscriber per month has declined in recent years, mobile text messaging traffic has increased.[8] About 88 percent of teenage mobile phone users now send and receive text messages, which is a rise from the 51 percent of teenagers who texted in 2006.[9]

Mobile phones are low-powered radio transceivers—a combination transmitter and receiver—that use radio waves to communicate with fixed installations, called base stations or cell towers. The radio waves used by mobile phones are a form of electromagnetic radiation—energy moving through space as a series of electric and magnetic waves. The spectrum of electromagnetic radiation comprises a range of frequencies from very low, such as electrical power from power lines, through visible light, to extremely high, such as gamma rays, as shown in figure 1. The portion of the electromagnetic spectrum used by mobile phones—as well as other telecommunications services, such as radio and television broadcasting— is referred to as the RF spectrum.

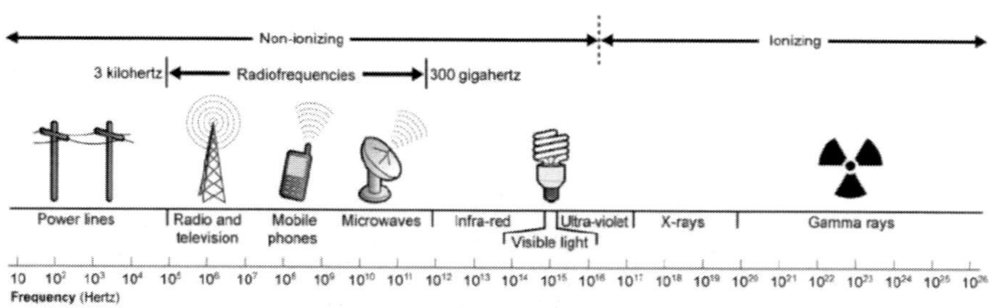

Source: FCC.

Figure 1. The Electromagnetic Spectrum.

The electromagnetic spectrum includes ionizing and non-ionizing radiation. Ionizing radiation, such as gamma rays, has energy levels high enough to strip electrons from atoms and molecules, which can lead to serious biological damage, including the production of cancers. RF energy, on the other hand, is in the non-ionizing portion of the electromagnetic spectrum, which lacks the energy needed to cause ionization. However, RF energy can produce other types of biological effects. For example, it has been known for many years that exposure to high levels of RF energy, particularly at microwave frequencies, can rapidly heat biological tissue. This thermal effect can cause harm by increasing body temperature, disrupting behavior, and damaging biological tissue. The thermal effect has been successfully harnessed for household and industrial applications, such as cooking food and molding plastics. Since mobile phones are required to operate at power levels well below the threshold for known thermal effects, the mobile phone health issue has generally focused on whether there are any adverse health effects from long-term or frequent exposure to low-power RF energy emissions that are not caused by heating.

RESEARCH ON RF ENERGY EXPOSURE FROM MOBILE PHONES HAS NOT DEMONSTRATED ADVERSE HEALTH EFFECTS, BUT MORE STUDIES ARE UNDER WAY

Scientific Research

Scientific research to date has not demonstrated adverse human health effects from RF energy exposure from mobile phone use, but additional research may increase understanding of possible effects. In 2001, we reported that FDA and others had concluded that research had not shown RF energy emissions from mobile phones to have adverse health effects, but that insufficient information was available to conclude mobile phones posed no risk.[10] Following another decade of scientific research and hundreds of studies examining health effects of RF energy exposure from mobile phone use, FDA maintains this conclusion. FDA stated that while the overall body of research has not demonstrated adverse health effects, some individual studies suggest possible effects. Officials from NIH, experts we interviewed, and a working group commissioned by IARC— the World Health Organization's agency that promotes international collaboration in cancer research—have reached similar conclusions. For example, in May 2011 IARC classified RF energy as "possibly carcinogenic to humans."[11] IARC determined that the evidence from the scientific research for gliomas, a type of cancerous brain tumor, was limited—meaning that an association has been observed between RF energy exposure and cancer for which a causal relationship is considered to be credible, but chance, bias, or confounding factors could not be ruled out with reasonable confidence.[12] With respect to other types of cancers, IARC determined that the evidence was inadequate—meaning that the available studies are of insufficient quality, consistency, or statistical power to permit a conclusion about the causal association. Additionally, in April 2012 an advisory group to the Health Protection Agency—an independent organization established by the United Kingdom government to protect the public from environmental hazards and infectious diseases—concluded that although there is substantial research on this topic, there is no convincing evidence that RF energy below guideline levels causes health effects in adults or children.[13]

A broad body of research is important for understanding the health effects of RF energy exposure from mobile phone use, because no single study can establish a cause-and-effect relationship and limitations associated with studies can make it difficult to draw conclusions. Two types of studies, epidemiological and laboratory, are used in combination to examine effects from mobile phones. Epidemiological studies investigate the association, if any, between health effects and the characteristics of people and their environment. Laboratory studies conducted on test subjects—including human volunteers, laboratory animals, biological tissue samples, or isolated cells—are used to determine a causal relationship between possible risk factors and human health, and the possible mechanisms through which that relationship occurs.

Epidemiological Studies

Studies we reviewed suggested and experts we interviewed stated that epidemiological research has not demonstrated adverse health effects from RF energy exposure from mobile phone use, but the research is not conclusive because findings from some studies have suggested a possible association with certain types of tumors, including cancerous tumors. Findings from one such study, the INTERPHONE study, were published in 2010.[14] This retrospective case-control study with more than 5,000 cases examined the association between mobile phone use and certain types of brain tumors, including cancerous tumors, in individuals aged 30-59 years in 13 countries.[15] Overall study findings did not show an increased risk of brain tumors from mobile phone use, but at the highest level of exposure, findings suggested a possible increased risk of glioma.[16] Other epidemiological studies have not found associations between mobile phone use and tumors, including cancerous tumors. For example, findings from a nationwide cohort study conducted in Denmark that originally followed 420,095 individuals did not show an association between increased risk for certain types of tumors, including cancerous tumors, and mobile phone use.[17] Additionally, findings from a subset of the cohort—56,648 individuals with 10 or more years since their first mobile phone subscription—did not show an increased risk for brain and nervous system tumors.[18] Further, these findings did not change for individuals in the cohort with 13 or more years since their first mobile phone subscription.[19] Also, the CEFALO study—an international case-control study that compared children aged 7 to 19 diagnosed with certain types of brain tumors, including brain cancers, to similar children who were not diagnosed with brain tumors—found no relationship between mobile phone use and risk for brain tumors.[20] Findings from another study, which was conducted by NIH and examined trends in brain cancer incidence rates in the United States using national cancer registry data collected from 1992 to 2006, did not find an increase in new cases of brain cancer, despite a dramatic increase in mobile phone use during this time period.[21]

Limitations associated with epidemiological studies can make it difficult to draw definitive conclusions about whether adverse health effects are linked to RF energy exposure from mobile phone use. One such limitation is that it is difficult to measure and control for all variables that may affect results. For example, it can be difficult to accurately measure RF energy exposure from mobile phone use because humans are exposed to RF energy from many sources within their environments and mobile phone technology and user patterns frequently change. Also, epidemiological studies to date have been limited in their ability to provide information about possible effects of long-term RF energy exposure because the prevalence of long-term mobile phone use is still relatively limited and some tumors, including some cancerous tumors, do not develop until many years after exposure. In addition, epidemiological studies, specifically cohort studies, are sometimes limited in their ability to provide information about increased risks for rare outcomes, such as certain types of brain tumors. To address challenges with assessing rare outcomes, case-control studies, which collect information about past mobile phone use among study participants, may be undertaken with large numbers of cases and controls. While these studies can potentially provide information on long-term use, and include enough cancer cases to examine whether this use is associated with rare diseases, collecting data in this way can introduce bias, such as recall bias, into study data and further limit findings. To mitigate this potential bias, some epidemiological studies, specifically cohort studies, follow large populations over time and collect data about mobile phone use before participants develop a certain outcome. In spite of these limitations, experts we spoke with told us that epidemiological studies are a key component of the body of research used for assessing the health effects of mobile phones.

Laboratory Studies

Studies we reviewed suggested and experts we interviewed stated that laboratory research has not demonstrated adverse human health effects from RF energy exposure from mobile phone use, but the research is not conclusive because findings from some studies have observed effects on test subjects. Some laboratory studies have examined whether RF energy has harmful effects by exposing samples of human and animal cells to RF energy over a range of dose rates, durations, and conditions to detect any changes in cellular structures and functions. For example, some studies have examined the effects of RF energy on deoxyribonucleic acid (DNA) in rodent and human cells. While some of these studies found that RF energy exposure damaged DNA,[22] others failed to replicate such an effect using similar experimental conditions.[23] Other studies have exposed laboratory animals to RF energy, examined the animals for changes, and compared outcomes with a control group. For example, some studies have measured the behavior or cognitive functioning of rats to assess the neurological effects of RF energy.[24] According to some studies we reviewed, while some of these studies have observed changes in behavior and cognitive function, overall, these studies have not consistently found adverse effects from RF energy levels emitted from mobile phones. Laboratory studies also have exposed human volunteers to RF energy to investigate possible effects, such as effects on the neurological system or blood pressure. According to studies we reviewed, some studies on human volunteers have observed changes, such as changes in brain activity, but the implications of these physiological changes in relation to adverse effects on human health are unknown.[25]

Limitations associated with laboratory studies can make it difficult to draw conclusions about adverse human health effects from RF energy exposure from mobile phone use. For example, studies conducted on laboratory animals allow researchers to examine the effects of RF energy exposure on animal systems, but this type of research is limited because effects on laboratory animals may not be the same on humans. Additionally, studies on test subjects may observe biological or physiological changes, but in some circumstances it is unclear how or even if these changes affect human health. Further, to increase the strength of the evidence that observed changes in laboratory studies are the effect of RF energy exposure, studies must be replicated and confirmed with additional research using different dose rates, durations, and conditions of RF energy while observing similar effects. To date, according to FDA officials and some experts we interviewed, only a few laboratory studies that have shown effects from RF energy have been replicated, and some replicated studies have not confirmed earlier results.

Areas for Additional Research

Studies we reviewed and experts we interviewed identified key areas for additional epidemiological and laboratory studies, and according to experts, additional research may increase understanding of any possible effects. For example, additional epidemiological studies, particularly large long-term prospective cohort studies and case-control studies on children, could increase knowledge on potential risks of cancer from mobile phone use. Also, studies and experts identified several areas for additional laboratory studies. For example, additional studies on laboratory animals as well as human and animal cells examining the possible toxic or harmful effects of RF energy exposure could increase knowledge on potential biological and health effects of RF energy. Further, additional laboratory studies on human and animal cells to examine non-thermal effects of RF energy could increase knowledge of how, if at all, RF energy interacts with biological systems. However, some experts we spoke to noted that, absent clear evidence for adverse health effects, it is difficult to justify investing significant resources in research examining non-thermal effects of RF energy from mobile phone use. Another area identified for additional laboratory research is studies on human volunteers examining the effect of changes in the neurological system, which could help determine if these possible observed changes in neurological functioning from RF energy are adverse effects. In addition to conducting additional research, experts we interviewed reported that the broader body of evidence on RF energy should be re-evaluated when findings from key studies become available, to determine whether additional research in certain areas is still warranted.

Current Research Activities

Current research activities of federal agencies, international organizations, and the mobile phone industry include funding and supporting ongoing research on the health effects of RF energy exposure from mobile phones. NIH is the only federal agency we interviewed that is directly funding ongoing studies on health effects of RF energy from mobile phone use. NIH officials reported that the agency has provided about $35 million for research in this area from 2001 to 2011. (See table 1 for more information on ongoing studies funded by NIH.)

Table 1. Ongoing NIH-Funded Studies on Health Effects of RF Energy Exposure from Mobile Phone Use

Description	NIH institute funding the study	Total NIH funding	Estimated year of completion
Examining environmental and genetic factors for meningioma, a type of brain tumor, at research sites in five states	National Cancer Institute	$8,779,998	2012
Evaluating brain cancer incidence trends in the United States using cancer registry data to determine if trends are consistent with reported epidemiological associations of mobile phone use and certain types of cancer	National Cancer Institute	Not applicable[a]	Not applicable[a]
Examining effects of mobile phones on brain glucose	National Institute on Alcohol Abuse and Alcoholism	$595,700	2012
Examining effects of exposure to mobile phones in childhood on the central nervous system using children in the Danish National Birth Cohort[b]	National Institute of Environmental Health Sciences	$423,500	2012
Examining toxicology and carcinogenic effects of RF energy in laboratory animals as part of the National Toxicology Program[c]	National Institute of Environmental Health Sciences	$25,600,000	2015

Source: GAO analysis of NIH information.

[a]The National Cancer Institute regularly monitors and evaluates the U.S. brain cancer incidence trends using Surveillance Epidemiology and End Result data. According to NIH officials, the National Cancer Institute does not separately track funding associated with performing this task. The most recent publication of data from this surveillance activity was published in 2012. See Little, M.P., et al, "Mobile Phone Use and Glioma Risk: Comparison of Epidemiological Study Results With Incident Trends In the United States," *British Medical Journal*, 2012, 344: e1147.

[b]The Danish National Birth Cohort consists of over 100,000 Danish children who were born from 1996 to 2002. Data on lifestyle factors, dietary habits, and environmental exposures have been collected on these children, and data on current mobile phone use by children have been collected since these children reached the age of seven.

[c]The National Toxicology Program is an interagency program that evaluates factors, such as RF energy, that could affect public health for the federal government. The three core federal agencies that make-up this program are NIH's National Institute of Environmental Health Sciences, CDC's National Institute for Occupational Safety and Health, and FDA's National Center for Toxicological Research. The National Toxicology Program is conducting comprehensive carcinogenicity studies on laboratory animals. According to FDA officials, FDA is conducting one of these National Toxicology Program studies in its National Center for Toxicological Research laboratory.

Although other federal agencies are not directly funding research in this area, some agencies are providing support for ongoing studies. For example, FDA officials reported that FDA's National Center for Toxicological Research, with funding provided by NIH as part of the National Toxicology

Program, is conducting studies on rat and bovine brain cells to examine whether RF energy emitted from mobile phones is toxic.[26] Also, CDC officials reported that the agency is collaborating with others to conduct ongoing studies in this area. For example, CDC officials reported that one of the agency's staff is collaborating with researchers in seven countries to conduct additional analyses on data collected through the INTERPHONE study to determine whether occupational exposure to RF energy and chemicals was a risk factor for brain cancer.

Federal agencies are also engaged in other activities to support research on the health effects of mobile phone use. For example, FDA collaborates with other organizations on research-related projects. According to FDA officials, the agency helped the World Health Organization develop its *WHO Research Agenda for Radiofrequency Fields* in 2001 and has provided comments to the World Health Organization on updates to this research agenda.[27] Also, officials from federal agencies that have responsibility for different aspects of RF energy safety and work—CDC, EPA, FCC, FDA, NIH, the National Telecommunications and Information Administration, and OSHA—are members of the Radiofrequency Interagency Work Group, which works to share information on RF energy related projects at the staff level. According to FCC and FDA officials, this group periodically meets to discuss RF energy related issues, including recently published and ongoing research on the health effects of RF energy exposure.

International organizations also support research on health effects of RF energy exposure from mobile phone use. Officials from IARC told us that the organization is currently supporting research activities for ongoing studies examining health effects of mobile phone use with respect to cancer. For example, IARC is involved in the identification of research sites for and implementation of the COSMOS study—a large international, prospective, cohort study that will follow individuals for 25 or more years to examine possible long-term health effects of using mobile phones, such as brain tumors, including cancers, and other health outcomes. IARC is also coordinating additional data analyses on previously published studies examining mobile phone health effects. For example, IARC is coordinating additional analyses of data collected for the INTERPHONE study. Additionally, the European Commission—the European Union's executive body that represents the interest of Europe as a whole—is supporting research in this field. Under its research program—the Seventh Framework Programme—the European Commission has provided funds for the MOBI-KIDS study, an international case-control study examining the possible association between communication technology, including mobile phones and other environmental exposures, and the risk of brain tumors in people aged 10 to 24 years.

The mobile phone industry supports research by providing funding for studies. According to representatives from mobile phone manufacturers, service providers, and industry associations, most industry funding for scientific research is provided by the Mobile Manufacturers Forum—an international not-for-profit association that is largely comprised of wireless device manufacturers. According to representatives from the Mobile Manufacturers Forum, the association has provided about $46 million for RF energy research since 2000 and is currently providing support for epidemiological and laboratory studies. Although representatives from all four mobile phone manufacturers that we interviewed reported that their companies support research through their industry associations, representatives from one manufacturer reported that it is also funding two studies examining the effects of RF energy emitted from mobile phones on human hands and the head.

FCC's RF Energy Exposure Limit May Not Reflect Latest Evidence on Thermal Effects, and Mobile Phone Testing Requirements May Not Identify Maximum Exposure

RF Energy Exposure Limit

In 1996, FCC adopted the RF energy exposure limit for mobile phones of 1.6 watts per kilogram, averaged over one gram of tissue, a measurement of the amount of RF energy absorbed into the body.[28] FCC developed its limit based on input from federal health and safety agencies as well as the 1991 recommendation by the Institute of Electrical and Electronics Engineers (IEEE) that was subsequently approved and issued in 1992 by the American National Standards Institute (ANSI).[29] This recommended limit was based on evidence related to the thermal effects of RF energy exposure[30]—the only proven health effects of RF energy exposure—and was set at a level well below the threshold for such effects. FCC noted that the limit provided a proper balance between protecting the public from exposure to potentially harmful RF energy and allowing industry to provide telecommunications services to the public in the most efficient and practical manner possible.

In 2006, IEEE published its updated recommendation for an RF energy exposure limit of 2.0 watts per kilogram, averaged over 10 grams of tissue. [31] This new recommended limit could allow for more RF energy exposure from mobile phone use, although actual exposure depends on a number of factors, including the operating power of the phone, how the phone is held during use, and where it is used in proximity to a mobile phone base station.[32] According to IEEE, improved RF energy research and a better understanding of the thermal effects of RF energy exposure on animals and humans, as well as a review of the available scientific research, led to the change in recommended RF energy exposure limit. IEEE's new recommended limit was harmonized with a 1998 recommendation of the International Commission on Non-Ionizing Radiation Protection, which has been adopted by more than 40 countries, including the European Union countries.[33] Both of these recommendations call for an exposure limit of 2.0 watts per kilogram averaged over 10 grams of tissue, which according to IEEE represents a scientific consensus on RF energy exposure limits.

According to senior FCC officials, the agency has not adopted any newer limit because federal health and safety agencies have not advised them to do so. FCC officials told us that they rely heavily on the guidance and recommendations of federal health and safety agencies when determining the appropriate RF energy exposure limit and that, to date, none of these agencies have advised FCC that its current RF energy limit needs to be revised. Officials from FDA and EPA told us that FCC has not formally asked either agency for an opinion on the RF energy limit. FDA officials noted, though, that if they had a concern with the current RF energy exposure limit, then they would bring it to the attention of FCC.

Although federal guidance states that agencies should generally use consensus standards, FCC officials provided reasons why they did not have current plans to change the RF energy exposure limit. Office of Management and Budget Circular A-119 concerning federal use of technical standards states that federal agencies must use "consensus standards in lieu of government-unique standards," except where inconsistent with law or otherwise impractical. FCC officials noted that no determination has been made that the new recommended RF energy exposure limit is inconsistent with law or impractical. FCC has recognized that research on RF energy exposure is ongoing and pledged to monitor the science to ensure that its guidelines continue to be appropriate.[34] FCC officials noted that an assessment of the current limit and the new recommended limit could be accomplished through a formal rulemaking process, which would include a solicitation of information and opinions from federal health and safety agencies.[35] FCC could alternatively release a Notice of Inquiry to gather information on this issue without formally initiating rulemaking.

Stakeholders we spoke with varied on whether the current U.S. RF energy exposure limit should be changed to reflect the new recommended limit. For instance, a few experts and consumer groups we spoke with said FCC should not adopt the new recommended exposure limit because of the relative uncertainty of scientific research on adverse health effects from mobile phone use. An official from one consumer group told us that adopting the 2.0 watts per kilogram exposure limit would be a step back, since it could allow users to be exposed to higher radiation levels. Conversely, some experts we spoke with maintained that both the 1.6- and 2.0-watts-per-kilogram limits protect users from the thermal effects of RF energy exposure—which the experts maintained are the only conclusively demonstrated effects of exposure—since a safety factor of fifty was applied to obtain the limits, meaning that the maximum permitted exposure is a fiftieth of what was determined to be the exposure at which potentially deleterious thermal effects are likely to occur.

Nevertheless, by not formally reassessing its current RF energy exposure limit, FCC cannot ensure that it is using a limit that reflects the latest evidence on thermal effects from RF energy exposure, and may impose additional costs on manufacturers and limitations on mobile phone design. FCC's current limit was established based on recommendations made more than 20 years ago. According to IEEE, the new recommended limit it developed is based on significantly improved RF research and therefore a better understanding of the thermal effects of RF energy exposure.

Additionally, three of the four mobile phone manufacturers we spoke with favored harmonization of RF energy exposure limits, telling us that maintaining the separate standards can result in additional costs and may affect phone design in a way that could limit performance and functionality. According to some manufacturers we spoke with, many of their phones are sold in multiple countries.

As a result, the manufacturers have to develop and test phones based on different exposure limits, which can require additional resources and slow the time it takes to get new phones into the market.

Additionally, one manufacturer indicated that some features are not enabled on phones sold in the United States that are available in other countries to comply with FCC's current limit. A reassessment by FCC would help it to determine if any changes to the limit are appropriate.

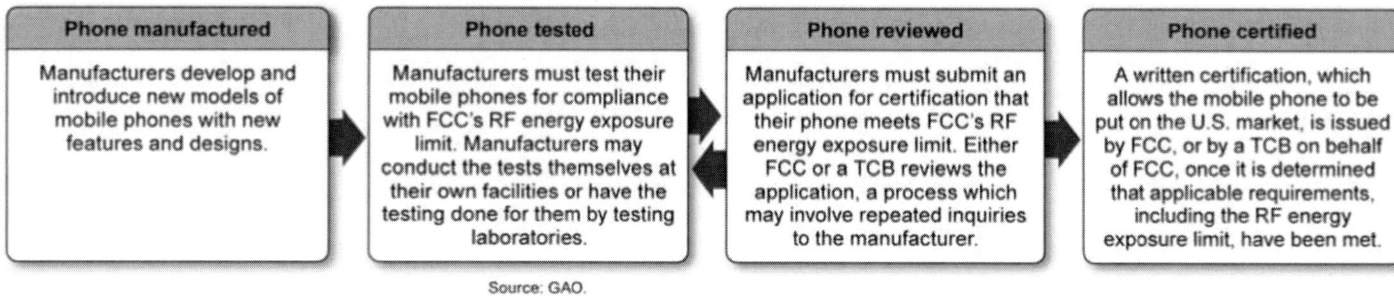

Source: GAO.

Figure 2. U.S. Mobile Phone Certification Process.

Mobile Phone Certification

FCC ensures compliance with its RF energy exposure limit by certifying all mobile phones sold in the United States. In its application for certification, manufacturers must provide evidence that their mobile phones meet FCC's RF energy exposure limit. FCC has authorized 23 TCBs in the United States and other countries to review applications that involve evaluation of RF exposure test data and issue certifications on behalf of the agency. TCBs are private organizations that have been accredited to perform these functions.[36] TCBs now perform the majority of mobile phone certifications, with FCC generally only handling the more complex certifications, such as mobile phones with multiple transmitters using third generation and fourth generation technology.[37] Figure 2 illustrates the mobile phone certification process.

Representatives from mobile phone manufacturers we spoke with were generally satisfied with how TCBs review and certify mobile phones, but noted that complex certifications handled by FCC can take a long time to process. For instance, since there are generally no established test procedures for new technologies, FCC must work with the manufacturer to develop appropriate procedures by which the agency can determine if the device meets the RF energy exposure limit. According to FCC, part of this review may result in changes to testing guidance. For example, representatives from one manufacturer told us that FCC may take many months to process an application for a newer product. FCC officials told us that over the last 10 years, the average time to review an application submitted directly to the agency has ranged from 45 to 60 days. Representatives from one TCB we spoke with noted that the TCB review can be as short as a week, though FCC does not collect data on how long it takes TCBs to process applications.

Mobile Phone Testing

To ensure that mobile phones comply with FCC's RF energy exposure limit, manufacturers conduct tests at their own laboratories or have the testing conducted for them by private laboratories. Laboratories must follow standardized FCC testing procedures or work with FCC to develop acceptable alternatives in some complex cases. These procedures require that the SAR be measured to ensure the mobile phone's compliance with the FCC exposure limit, which was designed to ensure that mobile phones do not expose the public to levels of RF energy that could be potentially harmful. FCC periodically updates the testing procedures as new mobile phone technology is introduced. A typical testing set-up is shown in figure 3.

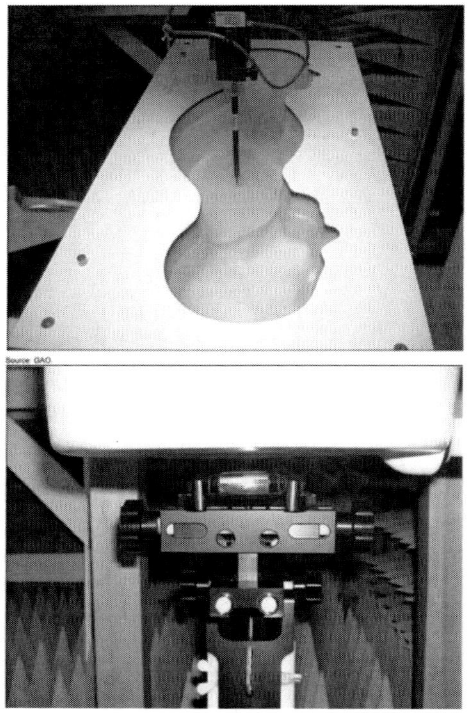

Source: GAO.

Note: To test mobile phones, a mold in the shape of an adult torso and head is filled with fluid mixture designed to simulate the electrical properties of human tissue. A phone is placed near the head or torso (the torso, or body, testing is illustrated above) and operated at maximum power. A probe attached to a computer-controlled mechanical arm is inserted into the mixture at various locations to measure SAR. This procedure is repeated for a number of closely specified phone positions and operating frequencies. To receive FCC certification, none of the SAR measurements can exceed FCC's exposure limit of 1.6 watts per kilogram.

Figure 3. Photographs of Mobile Phone Testing near the Body.

FCC has implemented standardized testing procedures requiring mobile phones to be tested for compliance with the RF energy exposure limit when in use against the ear and against the body while in body-worn accessories, such as holsters, but these requirements may not identify the maximum exposure under other conditions.[38] The specific minimum separation distance from the body is determined by the manufacturer (never to exceed 2.5 centimeters), based on the way in which the mobile phone is designed to be used.[39] The results of these testing requirements are two different values: a maximum SAR value for the head and a maximum SAR value for the body. However, these testing procedures may not identify the maximum SAR for the body, since some consumers use mobile phones with only a slight distance, or no distance, between the device and the body, such as placing the phone in a pocket while using an ear piece. Using a mobile phone in this manner could result in RF energy exposure above the maximum body-worn SAR determined during testing, although that may not necessarily be in excess of the FCC's limit. In such a case, exposure in excess of FCC's limit could occur if the device were to transmit continuously and at maximum power.

FCC has not reassessed its testing requirements to ensure that testing identifies the maximum RF energy exposure for the other usage conditions a user could experience when mobile phones are in use without body-worn accessories or as advised by the manufacturer's instructions, rather than the head. Although FCC officials said that they provide case-by-case guidance for many mobile phones operating with new technologies, they do not require testing of mobile phones when used without body-worn accessories unless such conditions are specifically identified by the manufacturer's operating instructions. Representatives of some consumer groups we spoke with expressed concern about the exposure to RF energy that can come with such use. Officials from IEEE, though, told us that the average power and resultant radiation level of mobile phones while in use is very low, such that even when a mobile phone is used against the body it is unlikely that the RF energy exposure would exceed the FCC limit. Nevertheless, FCC has not reassessed its testing requirements to ensure that mobile phones do not exceed the RF energy exposure limit in all possible usage conditions.

Beyond the testing required for certification, FCC also ensures that mobile phones meet its RF energy exposure limit by reviewing information collected as part of routine surveillance of mobile phones on the market. FCC requires TCBs to carry out this post-market surveillance program, through which each TCB tests one percent of the mobile phones they have certified for RF energy exposure, to ensure that the phones continue to meet FCC's RF energy exposure limit.[40] According to FCC, no mobile phone tested under this surveillance program has been found in violation of the RF energy exposure limit.

FEDERAL AGENCIES AND MOBILE PHONE INDUSTRY PROVIDE INFORMATION TO THE PUBLIC THROUGH WEBSITES AND USER MANUALS

Information Provided by Federal Agencies

Federal agencies provide information to the public on the health effects of mobile phone use and related issues primarily through their websites. This information includes summaries of research, and agencies' conclusions about the health effects of mobile phone use, as well as suggestions for how mobile phone users can reduce their exposure to RF energy. Table 2 summarizes selected information on mobile phones and health provided by six federal agencies on their websites.

Table 2. Federal Agency Website Information on Mobile Phones and Health as of June 2012

Types of information provided	Agency
What RF energy or radiation is	CDC, EPA, FCC, FDA, NIH, OSHA
Current mobile phone RF energy exposure limits	CDC, FCC, OSHA
How mobile phones are tested or certified	FCC
Thermal effects of RF energy exposure	CDC, FCC, NIH, OSHA
Non-thermal effects of RF energy exposure	CDC, FCC, OSHA
Health issues and research related to mobile phones	CDC, FCC, FDA, NIH, OSHA
Summaries or links to ongoing studies	CDC, FDA, NIH
Information on how to minimize or reduce RF energy exposure from mobile phone use	CDC, EPA, FCC, FDA, NIH, OSHA

Source: GAO analysis of federal agency websites.

Note: Some federal agency websites include additional information on mobile phones and health beyond the major topics listed above.

The types of information that federal agencies' websites provide on mobile phone health effects and related issues vary, in part because of the agencies' different missions, though the websites provide a broadly consistent message. For instance, NIH primarily provides information about the research on health effects of RF energy exposure from mobile phone use, while FCC provides information on how mobile phones are tested and certified. Nevertheless, the concluding statements about whether RF energy exposure from mobile phone use poses a risk to human health are generally consistent across selected federal agencies' websites that we reviewed, though the specific wording of these concluding statements varies.

Representatives from some consumer groups and experts we spoke with raised concerns that the information on federal agency websites about mobile phone health effects is not precautionary enough, among other things. In particular, these representatives and experts said that federal agencies should include stronger precautionary information about mobile phones because of the uncertain state of scientific research on mobile phone health effects as well as the fact that current testing requirements may not identify the maximum possible RF energy exposure. Representatives from one consumer group also said that federal agency websites should provide more consumer information, such as the impact of different mobile phone technologies on RF energy exposure. Officials from FCC and NIH maintained that the information on their websites reflects the latest scientific evidence and provides sufficient information for consumers concerned about potential health effects related to mobile phones.

Some consumer groups noted that they would like FCC to mention IARC's recent classification of RF energy exposure as "possibly carcinogenic" on FCC's website. FCC noted that it generally defers to the health and safety agencies for reporting on new research, though FCC's website did include information on the recent INTERPHONE study when we reviewed the site in June 2012. FCC does provide links to CDC, EPA, FDA, and other websites, some of which have information about the IARC's classification.[41] FDA notes on its website that the IARC classification means there is limited evidence showing RF carcinogenicity in humans and insufficient evidence of carcinogenicity in experimental animals.

Some local governments are taking steps to provide precautionary information to consumers. For example, the city of San Francisco has developed a Web page on mobile phone health issues, including steps to reduce RF energy exposure from mobile phone use, and has passed an ordinance requiring local mobile phone retailers to distribute a flyer on ways that consumers can reduce their exposure.[42]

Information Provided by Mobile Phone Industry

The mobile phone industry provides information to consumers on the health effects of mobile phone use and related issues through user manuals and websites. The information provided in user manuals by manufacturers is voluntary, as there are no federal requirements that manufacturers provide any specific information to consumers about the health effects of mobile phone use.[43] Most manuals we reviewed provide information about how the device was tested and certified, as well as the highest energy exposure measurement associated with the device. Some manufacturers also provide suggestions, often based on information from FDA, to consumers about how to minimize their exposure, among other things.

All manuals we reviewed, except one, include a statement that, when used on the body, as opposed to against the ear, a minimum distance between the body and the mobile phone should be maintained. These distances ranged from 1.5 to 2.5 centimeters.

Since all mobile phones are tested for RF energy exposure compliance at a distance from the body, as discussed previously in this report, these instructions are consistent with how the devices were tested and certified by FCC. Some consumer groups and experts we spoke with noted that consumers could be unaware of these instructions if they do not read the entire user manual.

CONCLUSION

FCC's current RF energy exposure limit for mobile phones, established in 1996, may not reflect the latest evidence on the thermal effects of RF energy exposure and may impose additional costs on manufacturers and limitations on mobile phone design. FCC regulates RF energy emitted from mobile phones and relies on federal health and safety agencies to help determine the appropriate RF energy exposure limit. However, FCC has not formally asked FDA or EPA for their assessment of the limit since 1996, during which time there have been significant improvements in RF energy research and therefore a better understanding of the thermal effects of RF energy exposure. This evidence has led to a new RF energy exposure limit recommendation from international organizations.

Additionally, maintaining the current U.S. limit may result in additional costs for manufacturers and impact phone design in a way that could limit performance and functionality. Reassessing its current RF energy exposure limit would ensure that FCC's limit protects the public from exposure to RF energy while allowing industry to provide telecommunications services in the most efficient and practical manner possible.

The current testing requirements for mobile phones may not identify the maximum RF energy exposure when tested against the body. FCC testing requirements state that mobile phone tests should be conducted with belt-clips and holsters attached to the phone or at a predetermined distance from the body.

These requirements were developed by FCC to identify the maximum RF energy exposure a user could experience when using a mobile phone, to ensure that the mobile phone meets the agency's RF energy exposure limit. This limit was designed to ensure that mobile phones do not expose the public to levels of RF energy that could be potentially harmful. By testing mobile phones only when at a distance from the body, FCC may not be identifying the maximum exposure, since some users may hold a mobile phone directly against the body while in use.

Using a mobile phone in this manner could result in RF energy exposure above the maximum body-worn SAR determined during testing, although that may not necessarily be in excess of FCC's limit. Reassessing its testing requirements would allow FCC to ensure that phones used by consumers in the United States do not result in RF energy exposure in excess of FCC's limit.

RECOMMENDATIONS FOR EXECUTIVE ACTION

We recommend that the Chairman of the FCC take the following two actions:

- Formally reassess the current RF energy exposure limit, including its effects on human health, the costs and benefits associated with keeping the current limit, and the opinions of relevant health and safety agencies, and change the limit if determined appropriate.
- Reassess whether mobile phone testing requirements result in the identification of maximum RF energy exposure in likely usage configurations, particularly when mobile phones are held against the body, and update testing requirements as appropriate.

AGENCY COMMENTS AND OUR EVALUATION

We provided a draft of this report to the Department of Commerce, Department of Defense, Department of Health and Human Services, Department of Labor, EPA, and FCC for review and comment. FCC provided comments in a letter from the Chief, Office of Engineering and Technology. In this letter, FCC noted that FCC's staff has independently arrived at the same conclusions about the RF exposure guidelines as GAO. FCC also noted that a draft Order and Further Notice of Proposed Rulemaking, along with a new Notice of Inquiry, which has been submitted by FCC staff to the Commission for their consideration, has the potential to address the recommendations made in this report. We agree that FCC's planned actions may address our recommendations. However, since FCC has not yet initiated a review of the RF energy exposure limit or mobile phone testing requirements, our recommendations are still relevant. FCC and the Departments of Commerce, Defense, and Health and Human Services also provided technical comments, which were incorporated as appropriate. The Department of Labor and EPA did not provide comments on the draft.

Mark L. Goldstein
Director, Physical Infrastructure

Marcia Crosse
Director, Health Care

APPENDIX I: SCOPE AND METHODOLOGY

To determine what is known about the human health effects of radio-frequency (RF) energy exposure from mobile phone use, we reviewed selected studies including studies and reports that review and assess the scientific research, such as meta-analyses and government reports, as well as key individual epidemiological and laboratory studies.[1] We identified 384 studies that examine the health effects of RF energy emitted from mobile phone use through literature searches and interviews. We conducted literature searches in six online databases with health and engineering content—Embase, Inspec, Medline, National Technical Information Service Bibliographic, SciSearch, and SocialSciSearch—containing peer-reviewed publications and government reports to identify studies published from January 2006 through September 2011 using health-, mobile phone-, and RF energy-related search terms. Additionally, we interviewed officials from federal agencies and representatives of academic institutions, consumer groups, and industry associations to identify studies published through December 2011. To select studies for our review, we conducted a preliminary review of the 384 studies and included those that met the following criteria: (1) reviewed and assessed the scientific research in a systematic way, such as meta-analyses, and discussed their methods for identifying, selecting, and assessing the scientific research that were used to draw conclusions or (2) were key reports that identify areas for additional research in these fields, such as the 2008 National Research Council's *Identification of Research Needs Relating to Potential Biological or Adverse Health Effects of Wireless Communication*.[2] We selected 38 studies that met these criteria. (See app. II for a list of the 38 studies we reviewed.)

To collect information on the 38 selected studies, we developed a data collection instrument that contained 16 open- and closed-ended questions about the entity or entities that published and funded the study; the study methods, key findings, and limitations; and additional research needs. To apply this data collection instrument, one analyst reviewed each study and recorded information in the data collection instrument. A second analyst then reviewed each completed data collection instrument to verify the accuracy of the information recorded. We summarized the findings and limitations of studies based on the completed data collection instruments, as well as areas for additional research identified in the studies. Additionally, we used this analysis to identify key, individual, epidemiological and laboratory studies.

We also interviewed subject matter experts to determine what is known about the human health effects of RF energy exposure from mobile phone use. First, we identified 123 potential subject matter experts to interview through the following sources: (1) interviews with officials from federal agencies and representatives of academic institutions, consumer groups, and industry associations and (2) participant lists of recent expert panels and workgroups on this topic. These panels and workgroups included:

- The National Research Council's Committee on Identification of Research Needs Relating to Potential Biological or Adverse Health Effects of Wireless Communications Devices,[3]
- The International Agency for Research on Cancer's (IARC) Monograph Working Group on RF electromagnetic fields,[4]

- The INTERPHONE Study Group,[5] and
- The European Commission's Scientific Committee on Emerging and Newly Identified Health Risks.[6]

Second, we assigned each expert to one or more broad categories that captured his or her general area of expertise. Next, we e-mailed those experts who, based on our initial review, (1) were identified through at least one source and we had information on their general area of expertise or (2) were identified through at least two sources regardless of whether we had information on their general area of expertise. We received responses from 42 experts agreeing to help us with our study. Based on these responses, we selected a judgmental sample of 11 experts who represented a range of expertise and professional backgrounds including public health and policy; biology and medicine; biostatistics; epidemiology; engineering, including bioelectrical engineering; and RF energy standards. (See table 3 for the list of individuals interviewed.) These experts were interviewed as individuals, not as representatives of any institution. Further, all of the experts completed a form stating that they had no conflicts of interest that would affect their ability to provide us with their perspectives on what is known about the human health effects of RF energy exposure from mobile phone use and related issues.

To determine the current research activities of federal agencies and other organizations related to mobile phone use and health, we interviewed representatives from various agencies and organizations. We identified agencies and organizations by reviewing information on their websites on RF energy and conducting interviews with officials from federal agencies and representatives of organizations familiar with research on health effects of mobile phone use. To determine the current research activities of federal agencies related to mobile phone use and health, we interviewed officials from the Department of Defense; Department of Health and Human Services' Centers for Disease Control and Prevention (CDC), Food and Drug Administration (FDA), and National Institutes of Health (NIH); Department of Labor's Occupational Safety and Health Administration (OSHA); Environmental Protection Agency (EPA); and Federal Communications Commission (FCC). To determine the research activities of other organizations, we interviewed representatives from IARC, academic institutions, consumer groups, mobile phone industry associations, mobile phone manufacturers, and mobile phone providers.

Table 3. Subject Matter Experts Interviewed

Name	Title	Institution[a]
Carl Blackman	a founder and former President (1990-91)	Bioelectromagnetics Society
Linda Erdreich	Senior Managing Scientist	Center for Epidemiology and Computational Biology, Exponent
Jukka Juutilainen	Professor of Radiation Biology and Radiation Epidemiology	Department of Environmental Science, University of EasternFinland
Leeka Kheifets	Professor of Epidemiology	Department of Epidemiology, School of Public Health, University of California, Los Angeles
Henry Lai	Research Professor	Department of Bioengineering, University of Washington
James Lin	Professor of Electrical Engineering, Bioengineering, Physiology, and Biophysics	University of Illinois, Chicago
David McCormick	Senior Vice President and Director	IIT Research Institute
Martin Röösli	Assistant Professor	Unit for Environmental Exposures and Health, Swiss Tropical and Public Health Institute, Basel
Siegal Sadetzki	Head Associate Professor	Cancer and Radiation Epidemiology Unit, The Gertner Institute, Chaim Sheba Medical Center, Israel Sackler School of Medicine, Tel-Aviv University, Israel
Jonathan Samet	Professor and Flora L. Thornton Chair	Department of Preventive Medicine, Keck School of Medicine, University of Southern California
Bernard Veyret	Senior Scientist	National Center for Scientific Research, Bordeaux

Source: GAO.

[a]We interviewed experts as individuals, not as representatives of any institution. We provide information on institutions to help readers identify experts.

To determine how FCC set the RF energy exposure limit and ensures compliance with it, we reviewed and summarized FCC regulations and guidance as well as reports from international organizations that recommend RF energy exposure limits. We also reviewed and summarized FCC testing and certification regulations and guidance for mobile phones. We conducted interviews with officials from FCC and representatives from selected Telecommunication Certification Bodies (TCBs). We selected the four TCBs that approved the most mobile phone certification applications for fiscal years 2000-2011 according to FCC: PCTEST Engineering Laboratory, Inc.; ACB, Inc.; CETECOM ICT Services GmbH; and Timco Engineering, Inc. These four TCBs have approved 69 percent of all U.S. mobile phone applications since 2000. We interviewed representatives from National Institute of Standards and Technology, American National Standards Institute, and American Association for Laboratory Accreditation to discuss their role in accrediting entities as TCBs and monitoring the activities of current TCBs. We also conducted interviews with representatives of the mobile phone industry and consumer groups for their perspectives on RF energy exposure limits as well as the testing and certification of mobile phones. Representatives of the mobile phone industry we spoke with included industry associations (CTIA-The Wireless Association and Mobile Manufacturers Forum) as well as the top four mobile phone service providers (AT&T, Sprint, T-Mobile, and Verizon) that represent about 90 percent of U.S. mobile phone service subscribers. We also spoke with representatives from four mobile phone manufacturers that represent over 70 percent of the U.S. market (LG, Motorola, Nokia, and Samsung).

To determine the actions federal agencies and the industry take to inform the public about issues related to mobile phone health effects, we reviewed the information on federal agency websites. We identified six federal agencies that have information about mobile phones and health-related issues on their websites: CDC, EPA, FCC, FDA, NIH, and OSHA. We conducted interviews with officials from those federal agencies to learn how they developed and update their websites. We spoke with representatives of the mobile phone industry noted above and consumer groups to obtain perspectives on the strengths and limitations of federal agency public-information-sharing efforts. We also spoke with the representatives of the mobile phone industry about how and why manufacturers include warnings or specific usage guidelines in their user manuals. Finally, we reviewed the user manuals of selected mobile phones (see table 4) to identify the usage and health information being provided to consumers, including any instructions to hold the mobile phone away from the body during use. The specific mobile phone models were identified by the manufacturers we spoke with as their top selling models in 2011.

Table 4. Mobile Phone User Manuals Reviewed

Manufacturer	Phone model
Apple[a]	iPhone 4
LG	Octane
	Optimus
Motorola	Bionic
	Razr
Nokia	1616
	6350
	X2-01
Samsung	GoPhone
	Gusto
	TracFone

Source: GAO.

[a]We included the Apple iPhone because of its prominence in the industry. Representatives from Apple declined to speak with us for this report.

APPENDIX II: STUDIES GAO REVIEWED

Ahlbom, Anders, Maria Feychting, Adele Green, Leeka Kheifets, David A. Savitz, and Anthony J. Swerdlow. "Epidemiological Evidence on Mobile Phones and Tumor Risk: A Review." *Epidemiology*, vol. 20, no. 5 (2009): 639-652.

Balbani, Aracy Pereira Silveira, and Jair Cortez Montovani. "Mobile Phones: Influence on Auditory and Vestibular Systems." *Brazilian Journal of Otorhinolaryngology*, vol 74, no. 1 (2008): 125-131.

Clapp, Richard W., Molly M. Jacobs, and Edward L. Loechler. "Environmental and Occupational Causes of Cancer: New Evidence 2005-2007." *Reviews on Environmental Health*, vol. 23, no. 1 (2008): 1- 37.

Committee on Man and Radiation. "COMAR Technical Information Statement: Expert Reviews on Potential Health Effects of Radiofrequency Electromagnetic Fields and Comments on the Bioinitiative Report." *Health Physics*, vol. 97, no. 4 (2009): 348-356.

Edumed Institute for Medicine and Health. *Non-Ionizing Electromagnetic Radiation in the Radiofrequency Spectrum and its Effects on Human Health.* Latin American Experts Committee on High Frequency Electromagnetic Fields and Human Health. June 2010.

European Health Risk Assessment Network on Electromagnetic Fields Exposure. "Risk Analysis of Human Exposure to Electromagnetic Fields." *Deliverable Report D2*, Executive Agency for Health and Consumers Framework of the Programme of Community Action in the Field Of Health 2008-2013. July 2010.

European Health Risk Assessment Network on Electromagnetic Fields Exposure. "D3 – Report on the analysis of risks associated to exposure to EMF: in vitro and in vivo (animals) studies." July 2010.

French Environmental Health and Safety Agency. "AFSSE Statement on Mobile Phones and Health." AFSSE. April 16, 2003.

German Mobile Telecommunication Research Programme. "Health Risk Assessment of Mobile Communications." Department Radiation Protection and Health. Germany: 2008.

Habash, Riadh W.Y., J. Mark Elwood, Daniel Krewski, W. Gregory Lotz, James P. McNamee, and Frank S. Prato. "Recent Advances in Research On Radiofrequency Fields and Health: 2004-2007." *Journal of Toxicology and Environmental Health*, Part B, vol. 12 (2009): 250-288.

Han, Yueh-Ying, Hideyuki Kano, Devra L. Davis, Ajay Niranjan, and L. Dade Lunsford. "Cell Phone Use and Acoustic Neuroma: The Need for Standardized Questionnaires and Access to Industry Data." *Surgical Neurology*, vol. 72 (2009): 216-222.

Health Council of the Netherlands. "Electromagnetic Fields: Annual Update 2008." The Hague: Health Council of the Netherlands, 2008; publication no. 2009/02.

HERMO. *Health Risk Assessment of Mobile Communications.* A Finnish Research Programme. Finland: 2007.

Institution of Engineering and Technology. "The Possible Harmful Biological Effects of Low-Level Electromagnetic Fields of Frequencies up to 300 GHz." 2010 Position Statement, Institution of Engineering and Technology. United Kingdom: 2010.

International Commission on Non-Ionizing Radiation Protection. *Exposure to high frequency electromagnetic fields, biological effects and health consequences* (100 kHz-300 GHz). Germany: 2009.

Juutilainen, Jukka, Anne Höytö, Timo Kumlin, and Jonne Naarala. "Review of Possible Modulation-Dependent Biological Effects of Radiofrequency Fields." *Bioelectromagnetics*, vol. 35 (2011): 511-534.

Khurana, Vini G., Charles Teo, Michael Kundi, Lennart Hardell, and Michael Carlberg. "Cell Phones and Brain Tumors: A Review Including the Long-Term Epidemiological Data." *Surgical Neurology*, vol. 72 (2009): 205-215.

Kohli, D., A. Sachdev, and H. Vats. "Cell Phones and Tumor: Still In No Man's Land." *Indian Journal of Cancer*, vol. 46, no. 1 (2009): 5-12.

Kundi, Michael. "The Controversy About a Possible Relationship Between Mobile Phone Use and Cancer." *Ciencia & Saude Coletiva*, vol. 15, no. 5 (2010): 2415-2430.

Levis, Angelo G., Nadia Minicuci, Paolo Ricci, Valerio Gennaro, and Spiridione Garbisa. "Mobile Phones and Head Tumours. The Discrepancies in Cause-Effect Relationships in the Epidemiological Studies – How Do They Arise?" *Environmental Health*, vol. 10, no. 59. (2011): 1-15.

Marino, Andrew A., and Simona Carrubba. "The Effects of Mobile-Phone Electromagnetic Fields of Brain Electrical Activity: A Critical Analysis of the Literature." *Electromagnetic Biology and Medicine*, vol. 28 (2009): 250-274.

McKinlay, A.F., S.G. Allen, R. Cox, P.J. Dimbylow, S.M. Mann, C.R. Muirhead, R.D. Saunders, Z.J. Sienkiewicz, J.W. Stather, and P.R. Wainwright. "Advice on Limiting Exposure to Electromagnetic Fields (0-300 GHz)." *Documents of the NRPB*, vol. 15, no. 2. National Radiological Protection Board, 2004.

Mobile Telecommunications and Health Research Programme. "*Report 2007*." United Kingdom: October 2007.

Myung, Seung-Kwon, Woong Ju, Diana D. McDonnell, Yeon Ji Lee, Gene Kazinets, Chih-Tao Cheng, and Joel M. Moskowitz. "Mobile Phone Use and Risk of Tumors: A Meta-Analysis" *Journal of Clinical Oncology*, vol. 27, no. 33 (2009): 5565-5572.

National Council on Radiation Protection and Measurements. "Biological Effects of Modulated Radiofrequency Fields." *NCRP Commentary*, no. 18. Bethesda, MD: 2003.

National Research Council. *Identification of Research Needs Relating to Potential Biological or Adverse Health Effects of Wireless Communication*. Washington, D.C.: 2008.

Nieden, Anja zur, Corrina Dietz, Thomas Eikmann, Jürgen Kiefer, and Caroline E.W. Herr. "Physicians Appeals On the Dangers of Mobile Communication – What Is the Evidence? Assessment of Public Health Data." *International Journal of Hygiene and Environmental Health*, vol. 212 (2009): 576-587.

Pourlis, Aris F., "Reproductive and Developmental Effects of EMF in Vertebrate Animal Models." *Pathophysiology*, vol. 16 (2009): 179-189.

Regel, Sabine J., and Peter Achermann. "Cognitive Performance Measures in Bioelectromagnetic Research – Critical Evaluation and Recommendations." *Environmental Health*, vol. 10, no. 10 (2011).

Röösli, Martin, and Kerstin Hug. "Wireless Communication Fields and Non-Specific Symptoms of Ill Health: A Literature Review." *Weiner Medizinische Wochenschrift*, vol. 161, no. 9-10 (2011): 240-250.

Scientific Committee on Emerging and Newly Identified Health Risks. *Health Effects of Exposure to EMF*. European Commission. January 19, 2009.

Sixth Framework Programme. "*EMF-NET: Effects of the Exposure to Electromagnetic Fields: From Science to Public Health and Safer Workplace*." WP2.2 Deliverable report D4bis: Effects on Reproduction and Development. Italy: 2007.

Swedish Radiation Protection Authority. *Recent Research on EMF and Health Risks Fifth Annual Report from SSI:s Independent Expert Group on Electromagnetic Fields*, 2007, Sweden: 2008.

Swedish Radiation Safety Authority. *Recent Research on EMF and Health Risk Seventh Annual Report from SSM:s Independent Expert Group on Electromagnetic Fields*, 2010. Sweden: 2010.

Valentini, E., G. Curcio, F. Moroni, M. Ferrara, L. De Gannaro, and M. Bertini. "Neurophysiological Effects of Mobile Phone Electromagnetic Fields on Humans: A Comprehensive Review." *Bioelectromagnetics*, vol. 28 (2007): 415-432.

Vanderstraeten, Jacques, and Luc Verschaeve. "Gene and Protein Expression Following Exposure to Radiofrequency Fields from Mobile Phones." Environmental Health Perspectives, vol. 116, no. 9 (2008): 1131-1135.

Vijayalaxmi, and Thomas J. Prihoda. "Genetic Damage in Mammalian Somatic Cells Exposed to Radiofrequency Radiation: A Meta-Analysis of Data from 63 Publications (1990-2005)." *Radiation Research*, vol.169 (2008): 561-574.

World Health Organization. *WHO Research Agenda for Radiofrequency Fields*. Switzerland: 2010.

End Notes

[1] In this report, we use the term "mobile phone" to refer to handheld (i.e., wireless) cellular telephones, including newer personal communication devices, such as "smart phones," that can deliver voice, data, and images.

[2] 47 C.F.R. § 1.1307(b)(2).

[3] 21 U.S.C. §§ 360hh—360ss.

[4] 21 U.S.C. § 360ll.

[5] GAO, *Telecommunications: Research and Regulatory Efforts on Mobile Phone Health Issues*, GAO-01-545 (Washington, D.C.: May 7, 2001).

[6] Federal Communications Commission, *Annual Report and Analysis of Competitive Market Conditions With Respect to Mobile Wireless, Including Commercial Mobile Services*, Fifteenth Report (June 27, 2011).

[7] *Annual Report and Analysis of Competitive Market Conditions With Respect to Mobile Wireless, Including Commercial Mobile Services* (2011).

[8] *Annual Report and Analysis of Competitive Market Conditions With Respect to Mobile Wireless, Including Commercial Mobile Services* (2011).

[9] Lenhart, A., et al, *Teens and Mobile Phones*, Pew Internet & American Life Project (April 20, 2010).

[10] GAO-01-545.

[11] IARC's classification of RF energy is based on conclusions of an IARC working group of more than 30 scientists from 14 countries who reviewed the scientific evidence on the exposure to RF energy from personal devices, such as mobile phones and other sources. IARC published a summary of this working group's findings, see Baan, R., et al, "Carcinogenicity of Radiofrequency Electromagnetic Fields," *The Lancet Oncology*, 2011, 12(7): 624-626. According to IARC officials, the IARC Monograph containing the complete assessments of the working group will be published in fall 2012. IARC has five groups for classifying factors: 1—carcinogenetic to humans, 2A—probably carcinogenic to humans, 2B—possibly carcinogenic to humans, 3—not classifiable as to its carcinogenicity to humans, and 4—probably not carcinogenic to humans. As of July 11, 2012, IARC had classified 952 factors, of which 779 have been classified in groups 2B or 3. Factors classified in the 2B group include coffee and gasoline.

[12] IARC also determined that the evidence from the scientific research was limited for acoustic neuromas, a type of non-cancerous brain tumor.

[13] Health Protection Agency, *Health Effects from Radiofrequency Electromagnetic Fields,* RCE-20 (2012).

[14] This study was conducted at 16 research sites. Several publications are available on study findings from the individual research sites. Findings discussed here are the primary study findings published by the INTERPHONE Study Group. See The INTERPHONE Study Group, "Brain Tumour Risk In Relation To Mobile Telephone Use: Results of the INTERPHONE International Case-Control Study," *International Journal of Epidemiology*, 2010, 39: 675-694.

[15] A "case-control" study is a study that compares individuals with a particular disease or outcome—cases—to individuals without that disease or outcome—controls—to investigate if the outcome is associated with exposure to a specific factor. Case-control studies are sometimes called retrospective studies, because the outcome occurred before the study began.

[16] RF energy exposure levels from mobile phone use were measured in terms of (1) the number of years since first use, (2) cumulative number of calls, and (3) cumulative duration of calls. Analysis of the relationship between RF energy exposure and risk of four types of tumors—tumors of the brain including glioma and meningioma, acoustic nerve, and parotid gland—were done using these three measures. Individuals that fell into the highest level of exposure are those that reported 1,640 or more cumulative lifetime hours of mobile phone use, which ranged from less than 1 year to more than 10 years of use.

[17] A "cohort" study is a study that follows a defined group of people—the cohort—over time. Outcomes of the people in subsets of the cohort are compared to examine if a particular intervention or factor is associated with a particular outcome. Cohort studies are sometimes called prospective studies, although they can be performed either retrospectively from historical records or prospectively.

[18] Schüz, J., et al, "Cellular Telephone Use and Cancer Risk: Update of a Nationwide Danish Cohort," *Journal of the National Cancer Institute,* 2006, 98(23):1707-1713.

[19] Frei, P., et al, "Use of Mobile Phones and Risk of Brain Tumours: Update of Danish Cohort Study," *British Medicine Journal,* 2011, 343: d6387.

[20] Aydin, D., et al, "Mobile Phone Use and Brain Tumors in Children and Adolescents: A Multicenter Case-Control Study," *Journal of the National Cancer Institute*, 2011, 103: 1-13.

[21] Inskip, P.D., Hoover, R.N., Devesa, S.S., "Brain Cancer Incidence Trends In Relation To Cellular Telephone Use In the United States," *Neuro-Oncology,* 2010, 12(11): 1147-1151.

[22] For example see Nikolova, T., et al, "Electromagnetic Fields Affect Transcript Levels of Apoptosis-Related Genes In Embryonic Stem Cell-Derived Neural Progenitor Cells," *The FASEB Journal*, 2005, 12: 1686-1688 and Diem, E., et al, "Non-Thermal DNA Breakage by Mobile-Phone Radiation (1800 MHz) In Human Fibroblasts and In Transformed GFSHR17 Rat Granulosa Cells In Vitro," *Mutation Research*, 2005, 583(2): 178-183.

[23] For example see Speit, G., Schütz, P., Hoffmann, H., "Genotoxic Effects of Exposure To Radiofrequency Electromagnetic Fields (RF-EMF) In Cultured Mammalian Cells Are Not Independently Reproducible," *Mutation Research*, 2007, 626(1-2): 42-47 and Hook, G.J., et al, "Measurement of DNA Damage and Apoptosis In Molt-4 Cells After In Vitro Exposure To Radiofrequency Radiation," *Radiation Research*, 2004, 161(2):193-200.

[24] For example see Kumlin T., et al, "Mobile Phone Radiation and the Developing Brain: Behavioral and Morphological Effects in Juvenile Rats," *Radiation Research*, 2007, 168(4): 471-479 and Nittby, H., et al, "Cognitive Impairment in Rats After Long-Term Exposure to GSM-900 Mobile Phone Radiation," *Bioelectromagnetics*, 2008, 29(3): 219- 232.

[25] For example see Volkow, N.D., et al, "Effects of Cell Phone Radiofrequency Signal Exposure on Brain Glucose Metabolism," *Journal of the American Medical Association*, 2011, 305(8): 808-813 and Regel, S.J., et al, "Pulsed Radio-Frequency Electromagnetic Fields: Dose-Dependent Effects on Sleep, the Sleep EEG and Cognitive Performance," *Journal of Sleep Research*, 2007, 16: 253-258.

[26] The National Toxicology Program is an interagency program that evaluates factors, such as RF energy, that could affect public health for the federal government. The three core federal agencies that make-up this program are NIH's National Institute of Environmental Health Sciences, CDC's National Institute for Occupational Safety and Health, and FDA's National Center for Toxicological Research. The National Toxicology Program is conducting comprehensive carcinogenicity studies on laboratory animals. Collectively these studies will provide information about potential human health effects of RF energy exposure.

[27] The World Health Organization most recently updated this research agenda in 2010. See The World Health Organization, *WHO Research Agenda for Radiofrequency Fields*, Geneva, Switzerland (2010).

[28] 61 Fed. Reg. 41017, August, 7, 1996. This measurement is called the specific absorption rate (SAR) and is the widely accepted measurement of RF energy absorbed into the body in watts per kilogram, averaged over an amount of tissue ranging from the entire body to one gram.

[29] See IEEE Std. C95.1-1991 and 47 CFR Sec. 2.1093(d)(2).

[30] In scientific tests, animals had adverse behavioral effects once they absorbed enough RF energy to increase their body temperature by 1 degree Celsius. IEEE incorporated a safety factor into its standards for general human exposure by setting them at one-fiftieth the exposure shown to cause adverse effects in animals. Because this limit is based on whole-body exposure, it was further adjusted to account for the fact that mobile phones expose only a part of the body to RF energy.

[31] See IEEE Std. C95.1-2005.

[32] The output power of a phone is variable, using the minimum necessary for successful communication, and at any time will be a function of distance to the nearest mobile phone antenna and the presence of obstructions.

[33] See International Commission on Non-Ionizing Radiation Protection, *Guidelines for Limiting Exposure to Time-Varying Electric, Magnetic, and Electromagnetic Fields (Up to 300 GHz)* (1998). The International Commission on Non-Ionizing Radiation Protection is an independent scientific organization that provides guidance and advice on the health hazards of non-ionizing radiation exposure. Its recommended exposure limit is for frequencies up to 10 gigahertz. The IEEE recommendation was made for frequencies between 100 kilohertz and 3 gigahertz.

[34] In re Guidelines for Evaluating the Environmental Effects of Radiofrequency Radiation, 11 FCC Rcd 15123, 15125 (1996).

[35] FCC's rulemaking process includes multiple steps as outlined by law, with several opportunities for public participation. FCC generally begins the process by releasing a Notice of Proposed Rulemaking and establishing a docket to gather information submitted by the public or developed within FCC to support the proposed rule. FCC analyzes information in the docket and drafts a final rule.

[36] In 1999, FCC established a TCB program and requested that the National Institute of Standards and Technology accredit entities to perform TCB functions. The National Institute of Standards and Technology determined, in accordance with its procedures and in consultation with the FCC, that it would recognize qualified accreditation bodies to accredit TCBs. Subsequently, the National Institute of Standards and Technology approved ANSI in May 2000 and the American Association for Laboratory Accreditation in April 2007 as accreditation organizations for TCBs. These accreditation bodies assess new and current TCBs to ensure they comply with relevant requirements.

[37] According to FCC, third and fourth generation mobile phone technologies allow consumers to access a variety of different services and functionalities, such as Web browsing, e-mail, access to application stores, video conference or chat, mapping and navigation systems, mobile commerce, and the downloading of content.

[38] These procedures were based on IEEE Std. 1528-2003. Because mobile phones are not tested when in use directly against the body, FCC recommends that mobile phone user manuals note that a minimum separation distance must be maintained between the user's body and the phone to comply with RF exposure limits.

[39] FCC guidance states that mobile phone body-worn tests should be conducted with belt-clips and holsters attached to the phone and positioned against the flat surface of the mold in normal use configurations. If the manufacturer does not supply these accessories, a predetermined distance from the back of the mobile device to the flat surface of the mold is recommended.

[40] Testing may be performed at either the TCB's testing facilities or at a subcontracted test facility.

[41] FCC's links to the EPA and FDA websites were not functional when we reviewed them in April 2012. After we provided our draft report to FCC these links were fixed and were functional as of July 2012.

[42] The ordinance has been challenged in federal court. The case is currently being considered by the U.S. Court of Appeals for the 9th Circuit.

[43] FCC does require a statement that the mobile phone complies with the agency's RF energy exposure limit, among other things. 47 CFR § 2.1077.

End Notes for Appendix I

[1] Epidemiological studies investigate the association, if any, between health effects and the characteristics of people and their environment. Laboratory studies conducted on test subjects—including human volunteers, laboratory animals, biological tissue samples, or isolated cells—are used to determine a causal relationship between possible risk factors and human health, and the possible mechanisms through which that relationship occurs.

[2] See National Research Council, *Identification of Research Needs Relating to Potential Biological or Adverse Health Effects of Wireless Communication* (Washington, D.C.: 2008).

[3] National Research Council, *Identification of Research Needs Relating to Potential Biological or Adverse Health Effects of Wireless Communication* (Washington, D.C.: 2008).

[4] Baan, R., et al, "Carcinogenicity of Radiofrequency Electromagnetic Fields," *Lancet Oncology*, 2011, 12(7): 624-626.

[5] The INTERPHONE study is a retrospective case-control study that examined effects of mobile phone use on certain types of brain cancers or tumors in more than 5,000 cases aged 30-59 years in 13 countries. See Cardis, E, et al, "Brain Tumor Risk in Relation to Mobile Telephone Use: Results of the INTERPHONE International Case-Control Study," *International Journal of Epidemiology*, 2010, 39: 675-694.

[6] European Commission, *Health Effects of Exposure to EMF*, 2009.

In: Cell Phone Use and Health Risks
Editors: C. Kempson and E. Rahm
ISBN: 978-1-62257-948-8

Chapter 2

CELL PHONES AND CANCER RISK: FACTSHEET*

National Cancer Institute

Key Points

- Cell phones emit radiofrequency energy, a form of non-ionizing electromagnetic radiation, which can be absorbed by tissues closest to where the phone is held.
- The amount of radiofrequency energy a cell phone user is exposed to depends on the technology of the phone, the distance between the phone's antenna and the user, the extent and type of use, and the user's distance from cell phone towers.
- Studies thus far have not shown a consistent link between cell phone use and cancers of the brain, nerves, or other tissues of the head or neck. More research is needed because cell phone technology and how people use cell phones have been changing rapidly.

---

* This is an edited, reformatted and augmented version of U.S. Department of Health and Human Services, dated 6/18/2012.

1. Why Is There Concern That Cell Phones May Cause Cancer or Other Health Problems?

There are three main reasons why people are concerned that cell phones (also known as "wireless" or "mobile" telephones) might have the potential to cause certain types of cancer or other health problems:

- Cell phones emit radiofrequency energy (radio waves), a form of non-ionizing radiation. Tissues nearest to where the phone is held can absorb this energy.
- The number of cell phone users has increased rapidly. As of 2010, there were more than 303 million subscribers to cell phone service in the United States, according to the Cellular Telecommunications and Internet Association. This is a nearly threefold increase from the 110 million users in 2000. Globally, the number of cell phone subscriptions is estimated by the International Telecommunications Union to be 5 billion.
- Over time, the number of cell phone calls per day, the length of each call, and the amount of time people use cell phones have increased. Cell phone technology has also undergone substantial changes.

2. What Is Radiofrequency Energy and How Does It Affect the Body?

Radiofrequency energy is a form of electromagnetic radiation. Electromagnetic radiation can be categorized into two types: ionizing (e.g., x-rays, radon, and cosmic rays) and non-ionizing (e.g., radiofrequency and extremely low-frequency or power frequency).

Exposure to ionizing radiation, such as from radiation therapy, is known to increase the risk of cancer. However, although many studies have examined the potential health effects of non-ionizing radiation from radar, microwave ovens, and other sources, there is currently no consistent evidence that non-ionizing radiation increases cancer risk [1] .

The only known biological effect of radiofrequency energy is heating. The ability of microwave ovens to heat food is one example of this effect of radiofrequency energy. Radiofrequency exposure from cell phone use does cause heating; however, it is not sufficient to measurably increase body temperature.

A recent study showed that when people used a cell phone for 50 minutes, brain tissues on the same side of the head as the phone's antenna metabolized more glucose than did tissues on the opposite side of the brain [2]. The researchers noted that the results are preliminary, and possible health outcomes from this increase in glucose metabolism are still unknown.

3. How Is Radiofrequency Energy Exposure Measured in Epidemiologic Studies?

Levels of radiofrequency exposure are indirectly estimated using information from interviews or questionnaires. These measures include the following:

- How "regularly" study participants use cell phones (the minimum number of calls per week or month)
- The age and the year when study participants first used a cell phone and the age and the year of last use (allows calculation of the duration of use and time since the start of use)
- The average number of cell phone calls per day, week, or month (frequency)
- The average length of a typical cell phone call
- The total hours of lifetime use, calculated from the length of typical call times, the frequency of use, and the duration of use

4. What Has Research Shown About the Possible Cancer-Causing Effects of Radiofrequency Energy?

Although there have been some concerns that radiofrequency energy from cell phones held closely to the head may affect the brain and other tissues, to date there is no evidence from studies of cells, animals, or humans that radiofrequency energy can cause cancer.

It is generally accepted that damage to DNA is necessary for cancer to develop. However, radiofrequency energy, unlike ionizing radiation, does not cause DNA damage in cells, and it has not been found to cause cancer in animals or to enhance the cancer-causing effects of known chemical carcinogens in animals [3-5] .

Researchers have carried out several types of epidemiologic studies to investigate the possibility of a relationship between cell phone use and the risk of malignant (cancerous) brain tumors, such as gliomas, as well as benign (noncancerous) tumors, such as acoustic neuromas (tumors in the cells of the nerve responsible for hearing), most meningiomas (tumors in the meninges, membranes that cover and protect the brain and spinal cord), and parotid gland tumors (tumors in the salivary glands) [6].

In one type of study, called a case-control study, cell phone use is compared between people with these types of tumors and people without them. In another type of study, called a cohort study, a large group of people is followed over time and the rate of these tumors in people who did and didn't use cell phones is compared. Cancer incidence data can also be analyzed over time to see if the rates of cancer changed in large populations during the time that cell phone use increased dramatically. The results of these studies have generally not provided clear evidence of a relationship between cell phone use and cancer, but there have been some statistically significant findings in certain subgroups of people.

Findings from specific research studies are summarized below:

- The Interphone Study, conducted by a consortium of researchers from 13 countries, is the largest health-related case-control study of use of cell phones and head and neck tumors. Most published analyses from this study have shown no statistically significant increases in brain or central nervous system cancers related to higher amounts of cell phone use. One recent analysis showed a statistically significant, albeit modest, increase in the risk of glioma among the small proportion of study participants who spent the most total time on cell phone calls. However, the researchers considered this finding inconclusive because they felt that the amount of use reported by some respondents was unlikely and because the participants who reported lower levels of use appeared to have a reduced risk of brain cancer [7-9]. Another recent study from the group found no relationship between brain tumor locations and regions of the brain that were exposed to the highest level of radiofrequency energy from cell phones [10].

- A cohort study in Denmark linked billing information from more than 358,000 cell phone subscribers with brain tumor incidence data from the Danish Cancer Registry. The analyses found no association between cell phone use and the incidence of glioma, meningioma, or acoustic neuroma, even among people who had been cell phone subscribers for 13 or more years [11-13].
- Early case-control studies in the United States, Europe, and Japan were unable to demonstrate a relationship between cell phone use and glioma or meningioma [14].
- Some case-control studies in Sweden found statistically significant trends of increasing brain cancer risk for the total amount of cell phone use and the years of use among people who began using cell phones before age 20 [15]. However, another large, case-control study in Sweden did not find an increased risk of brain cancer among people between the ages of 20 and 69 [16]. In addition, the international CEFALO study, which compared children who were diagnosed with brain cancer between ages 7 and 19 with similar children who were not, found no relationship between their cell phone use and risk for brain cancer [17].
- NCI's Surveillance, Epidemiology, and End Results (SEER) Program, which tracks cancer incidence in the United States over time, found no increase in the incidence of brain or other central nervous system cancers between 1987 and 2008, despite the dramatic increase in cell phone use in this country during that time [18, 19]. Similarly, incidence data from Denmark, Finland, Norway, and Sweden for the period 1974–2008 revealed no increase in age-adjusted incidence of brain tumors [20, 21]. A 2012 study by NCI researchers, which compared observed glioma incidence rates in SEER with projected rates based on risks observed in the Interphone study [8], found that the projected rates were consistent with observed U.S. rates. The researchers also compared the SEER rates with projected rates based on a Swedish study published in 2011 [15]. They determined that the projected rates were at least 40 percent higher than, and incompatible with, the U.S. rates.
- Studies of workers exposed to radiofrequency energy have shown no evidence of increased risk of brain tumors among U.S. Navy electronics technicians, aviation technicians, or fire control technicians, those working in an electromagnetic pulse test program,

plastic-ware workers, cellular phone manufacturing workers, or Navy personnel with a high probability of exposure to radar [6].

5. Why Are the Findings From Different Studies of Cell Phone Use and Cancer Risk Inconsistent?

A limited number of studies have shown some evidence of statistical association of cell phone use and brain tumor risks, but most studies have found no association. Reasons for these discrepancies include the following:

- **Recall bias**, which may happen when a study collects data about prior habits and exposures using questionnaires administered after disease has been diagnosed in some of the study participants. It is possible that study participants who have brain tumors may remember their cell phone use differently than individuals without brain tumors. Many epidemiologic studies of cell phone use and brain cancer risk lack verifiable data about the total amount of cell phone use over time. In addition, people who develop a brain tumor may have a tendency to recall using their cell phone mostly on the same side of their head where the tumor was found, regardless of whether they actually used their phone on that side of their head a lot or only a little.
- **Inaccurate reporting**, which may happen when people say that something has happened more or less often than it actually did. People may not remember how much they used cell phones in a given time period.
- **Morbidity and mortality** among study participants who have brain cancer. Gliomas are particularly difficult to study, for example, because of their high death rate and the short survival of people who develop these tumors. Patients who survive initial treatment are often impaired, which may affect their responses to questions. Furthermore, for people who have died, next-of-kin are often less familiar with the cell phone use patterns of their deceased family member and may not accurately describe their patterns of use to an interviewer.
- **Participation bias**, which can happen when people who are diagnosed with brain tumors are more likely than healthy people (known as controls) to enroll in a research study. Also, controls who did not or rarely used cell phones were less likely to participate in the

Interphone study than controls who used cell phones regularly. For example, the Interphone study reported participation rates of 78 percent for meningioma patients (range 56–92 percent for the individual studies), 64 percent for the glioma patients (range 36–92 percent), and 53 percent for control subjects (range 42–74 percent) [9]. One series of Swedish studies reported participation rates of 85 percent in people with brain cancer and 84 percent in control subjects [16].

- **Changing technology and methods of use**. Older studies evaluated radiofrequency energy exposure from analog cell phones. However, most cell phones today use digital technology, which operates at a different frequency and a lower power level than analog phones. Digital cell phones have been in use for more than a decade in the United States, and cellular technology continues to change [6]. Texting, for example, has become a popular way of using a cell phone to communicate that does not require bringing the phone close to the head. Furthermore, the use of hands-free technology, such as wired and wireless headsets, is increasing and may decrease radiofrequency energy exposure to the head and brain.

6. What Do Expert Organizations Conclude?

The International Agency for Research on Cancer (IARC) (http://www.iarc.fr/en/media-centre/pr/2011/pdfs/pr208 E.pdf), a component of the World Health Organization, has recently classified radiofrequency fields as "possibly carcinogenic to humans," based on limited evidence from human studies, limited evidence from studies of radiofrequency energy and cancer in rodents, and weak mechanistic evidence (from studies of genotoxicity, effects on immune system function, gene and protein expression, cell signaling, oxidative stress, and apoptosis, along with studies of the possible effects of radiofrequency energy on the blood-brain barrier).

The American Cancer Society (ACS) (http://pressroom.cancer.org/ index.php?s=43&item=312) states that the IARC classification means that there could be some risk associated with cancer, but the evidence is not strong enough to be considered causal and needs to be investigated further. Individuals who are concerned about radiofrequency exposure can limit their exposure, including using an ear piece and limiting cell phone use, particularly among children.

The National Institute of Environmental Health Sciences (NIEHS) (http://www.niehs.nih.gov/health/topics/agents/cellphones/index.cfm) states that the weight of the current scientific evidence has not conclusively linked cell phone use with any adverse health problems, but more research is needed.

The U.S. Food and Drug Administration (FDA) (http://www.fda.gov/Radiation-EmittingProducts/RadiationEmittingProductsandProcedures/HomeBusinessandEntertainment/CellPhones/default.htm), which is responsible for regulating the safety of machines and devices that emit radiation (including cell phones), notes that studies reporting biological changes associated with radiofrequency energy have failed to be replicated and that the majority of human epidemiologic studies have failed to show a relationship between exposure to radiofrequency energy from cell phones and health problems.

The U.S. Centers for Disease Control and Prevention (CDC) (http://www.cdc.gov/nceh/radiation/cell phones. FAQ.html) states that, although some studies have raised concerns about the possible risks of cell phone use, scientific research as a whole does not support a statistically significant association between cell phone use and health effects.

The Federal Communications Commission (FCC) (http://www.fcc.gov/topic/wireless) concludes that there is no scientific evidence that proves that wireless phone use can lead to cancer or to other health problems, including headaches, dizziness, or memory loss.

7. What Studies Are Under Way That Will Help Further Our Understanding of the Health Effects of Cell Phone Use?

A large prospective cohort study of cell phone use and its possible long-term health effects was launched in Europe in March 2010. This study, known as COSMOS, will enroll approximately 250,000 cell phone users ages 18 or older and will follow them for 20 to 30 years.

Participants in COSMOS will complete a questionnaire about their health, lifestyle, and current and past cell phone use. This information will be supplemented with information from health records and cell phone records. More information about the COSMOS study is available at http://www.ukcosmos.org/index.html.

The challenge of such ambitious studies is to maintain the completeness of their cohorts over many decades. Researchers will need to determine whether participants who leave are somehow different from those who remain throughout the follow-up period.

Another case-control study, called Mobi-Kids, is under way to examine health effects among children. More information about the Mobi-Kids study is available at http://www.mbkds.net/.

Although recall bias is minimized in case-control studies that link to cell phone records, such studies face other problems. For example, it is impossible to know who is using the listed cell phone or whether that individual also places calls using other cell phones. To a lesser extent, it is not clear whether multiple users of a single phone will be represented on a single bill.

The NIEHS, which is part of the National Institutes of Health, is carrying out a study of risks related to exposure to radiofrequency energy (the type used in cell phones) in highly specialized labs that can specify and control sources of radiation and measure their effects on rodents.

8. Do Children Have a Higher Risk of Developing Cancer due to Cell Phone Use Than Adults?

In theory, children have the potential to be at greater risk than adults for developing brain cancer from cell phones. Their nervous systems are still developing and therefore more vulnerable to factors that may cause cancer. Their heads are smaller than those of adults and therefore have a greater proportional exposure to the field of radiofrequency radiation that is emitted by cell phones. And children have the potential of accumulating more years of cell phone exposure than adults do.

So far, the data from clinical studies in children do not support this theory. The first published analysis came from a large case-control study called CEFALO, which was conducted in Denmark, Sweden, Norway, and Switzerland. The study included children who were diagnosed with brain tumors between 2004 and 2008, when their ages ranged from 7 to 19. Researchers did not find an association between cell phone use and brain tumor risk in this group of children. However, they noted that their results did not rule out the possibility of a slight increase in brain cancer risk among children who use cell phones, and that data gathered through prospective studies and objective measurements, rather than participant surveys and recollections, will be key in clarifying whether there is an increased risk [18].

Researchers from the Centre for Research in Environmental Epidemiology in Spain are conducting another international study—Mobi-Kids—to evaluate the risk associated with new communications technologies (including cell phones) and other environmental factors in young people ages 10 to 24. More information about the Mobi-Kids study is available at http://www.mbkds.net/.

9. What Can Cell Phone Users Do to Reduce Their Exposure to Radiofrequency Energy?

The FDA and FCC have suggested some steps that concerned cell phone users can take to reduce their exposure to radiofrequency energy (1, 22):

- Reserve the use of cell phones for shorter conversations or for times when a landline phone is not available.
- Use a hands-free device, which places more distance between the phone and the head of the user.

Hands-free kits reduce the amount of radiofrequency energy exposure to the head because the antenna, which is the source of energy, is not placed against the head.

10. Where Can I Find More Information About Radiofrequency Energy from My Cell Phone?

The FCC provides information about the specific absorption rate (SAR) of cell phones produced and marketed within the last 1 to 2 years. The SAR corresponds with the relative amount of radiofrequency energy absorbed by the head of a cell phone user [23].

Consumers can access this information using the phone's FCC ID number, which is usually located on the case of the phone, and the FCC's ID search form, which is located at http://transition.fcc.gov/oet/ea/fccid/.

11. What Are Other Sources of Radiofrequency Energy?

The most common exposures to radiofrequency energy are from telecommunications devices and equipment [1]. In the United States, cell phones currently operate in a frequency range of about 1,800 to 2,200 megahertz (MHz) [6]. In this range, the electromagnetic radiation produced is in the form of non-ionizing radiofrequency energy.

Cordless phones (phones that have a base unit connected to the telephone wiring in a house) often operate at radio frequencies similar to those of cell phones; however, since cordless phones have a limited range and require a nearby base, their signals are generally much less powerful than those of cell phones.

Among other radiofrequency energy sources, AM/FM radios and VHF/UHF televisions operate at lower radio frequencies than cell phones, whereas sources such as radar, satellite stations, magnetic resonance imaging (MRI) devices, industrial equipment, and microwave ovens operate at somewhat higher radio frequencies [1] .

12. How Common Is Brain Cancer? Has the Incidence of Brain Cancer Changed over Time?

Brain cancer incidence and mortality (death) rates have changed little in the past decade. In the United States, 22,910 new diagnoses and 13,700 deaths from brain cancer are estimated for 2012.

The 5-year relative survival for brain cancers diagnosed from 2002 through 2008 was 35 percent [24] . This is the percentage of people diagnosed with brain cancer who will still be alive 5 years after diagnosis compared with the survival of a person of the same age and sex who does not have cancer.

The risk of developing brain cancer increases with age. From 2005 through 2009, there were fewer than 5 brain cancer cases for every 100,000 people in the United States under age 65, compared with approximately 19 cases for every 100,000 people in the United States who were ages 65 or older [24] .

REFERENCES

[1] U.S. Food and Drug Administration (2009). Radiation-Emitting Products: Reducing Exposure: Hands-free Kits and Other Accessories. Retrieved June 18, 2012, from: http://www.fda.gov/Radiation-EmittingProducts/RadiationEmittingProductsandProcedures/HomeBusinessandEntertainment/CellPhones/ucm1 16293.htm.

[2] Volkow ND, Tomasi D, Wang GJ, et al. Effects of cell phone radiofrequency signal exposure on brain glucose metabolism. *JAMA* 2011;305(8):808-813. [PubMed Abstract]

[3] Hirose H, Suhara T, Kaji N, et al. Mobile phone base station radiation does not affect neoplastic transformation in BALB/3T3 cells. *Bioelectromagnetics* 2008;29(1):55-64. [PubMed Abstract]

[4] Oberto G, Rolfo K, Yu P, et al. Carcinogenicity study of 217 Hz pulsed 900 MHz electromagnetic fields in Pim1 transgenic mice. *Radiation Research* 2007;168(3):316-326. [PubMed Abstract]

[5] Zook BC, Simmens SJ. The effects of pulsed 860 MHz radiofrequency radiation on the promotion of neurogenic tumors in rats. *Radiation Research* 2006;165(5):608-615. [PubMed Abstract]

[6] Ahlbom A, Green A, Kheifets L, et al. Epidemiology of health effects of radiofrequency exposure. *Environmental Health Perspectives* 2004;112(17):1741-1754. [PubMed Abstract]

[7] Cardis E, Richardson L, Deltour I, et al. The INTERPHONE study: design, epidemiological methods, and description of the study population. *European Journal of Epidemiolpgy* 2007;22(9):647-664. [PubMed Abstract]

[8] International Agency for Research on Cancer (2008). INTERPHONE Study: latest results update—8 October 2008. Lyon, France.Retrieved June 18, 2012, from: http://www.iarc.fr/en/researchgroups/RAD/Interphone8oct08.pdf.

[9] The INTERPHONE Study Group. Brain tumour risk in relation to mobile telephone use: results of the INTERPHONE international case-control study. *International Journal of Epidemiology* 2010;39(3):675-694. [PubMed Abstract]

[10] Larjavaara S, Schüz J, Swerdlow A, et al. Location of gliomas in relation to mobile telephone use: a case-case and case-specular analysis. *American Journal of Epidemiology* 2011;174(1):2-11. [PubMed Abstract]

[11] Johansen C, Boice J, McLaughlin J, Olsen J. Cellular telephones and cancer--a nationwide cohort study in Denmark. *Journal of the National Cancer Institute* 2001;93(3):203-207. [PubMed Abstract]

[12] Schüz J, Jacobsen R, Olsen JH, et al. Cellular telephone use and cancer risk: update of a nationwide Danish cohort. *Journal of the National Cancer Institute* 2006;98(23):1707-1713. [PubMed Abstract]

[13] Frei P, Poulsen AH, Johansen C, et al. Use of mobile phones and risk of brain tumours: update of Danish cohort study. *British Medical Journal* 2011;343:d6387. [PubMed Abstract]

[14] Muscat JE, Malkin MG, Thompson S, et al. Handheld cellular telephone use and risk of brain cancer. *JAMA* 2000;284(23):3001-3007. [PubMed Abstract]

[15] Hardell L, Carlberg M, Hansson Mild K. Pooled analysis of case-control studies on malignant brain tumours and the use of mobile and cordless phones including living and deceased subjects. *International Journal of Oncology* 2011;38(5):1465-1474. [PubMed Abstract]

[16] Lönn S, Ahlbom A, Hall P, Feychting M. Long-term mobile phone use and brain tumor risk. *American Journal of Epidemiology* 2005;161(6):526-535. [PubMed Abstract]

[17] Aydin D, Feychting M, Schüz J, et al. Mobile phone use and brain tumors in children and adolescents: a multicenter case-control study. *Journal of the National Cancer Institute* 2011;103(16):1264-1276. [PubMed Abstract]

[18] Inskip PD, Hoover RN, Devesa SS. Brain cancer incidence trends in relation to cellular telephone use in the United States. *Neuro-Oncology* 2010;12(11):1147-1151. [PubMed Abstract]

[19] Little MP, Rajaraman P, Curtis RE, et al. Mobile phone use and glioma risk: comparison of epidemiological study results with incidence trends in the United States. *British Medical Journal* 2012;344:e1147. [PubMed Abstract]

[20] Deltour I, Johansen C, Auvinen A, et al. Time trends in brain tumor incidence rates in Denmark, Finland, Norway, and Sweden, 1974-2003. *Journal of the National Cancer Institute* 2009;101(24):1721-1724. [PubMed Abstract]

[21] Deltour I, Auvinen A, Feychting M, et al. Mobile phone use and incidence of glioma in the Nordic countries 1979-2008: consistency check. *Epidemiology* 2012;23(2):301-307. [PubMed Abstract]

[22] U.S. Federal Communications Commission (2010) Wireless. Washington, D.C. Retrieved June 18, 2012, from: http://www.fcc.gov/topic/wireless.

[23] U.S. Federal Communications Commission (n.d.). FCC Encyclopedia: Specific Absorption Rate (SAR) for Cellular Telephones. Retrieved June 18, 2012, from: http://www.fcc.gov/encyclopedia/specific-absorption-rate-sar-cellular-telephones.

[24] Howlader N, Noone AM, Krapcho M, et al. (2012). *SEER Cancer Statistics Review, 1975–2009 (Vintage 2009 populations).* Bethesda, MD: National Cancer Institute. Retrieved June 18, 2012, from: http://seer.cancer.gov/csr/19752009pops09/index.html.

Related Resources

- Magnetic Field Exposure and Cancer (http://www.cancer.gov/cancertopics/factsheet/Risk/magnetic-fields)
- Cancer Causes and Risk Factors Home Page (http://www.cancer.gov/cancertopics/causes)

In: Cell Phone Use and Health Risks
Editors: C. Kempson and E. Rahm
ISBN: 978-1-62257-948-8

Chapter 3

Statement of Dariusz Leszczynski, Ph.D., Research Professor, Radiation and Nuclear Safety Authority, University of Helsinki. Hearing on "Health Effects of Cell Phone Use"*

Brief Biographical Sketch

My name is Dariusz Leszczynski and I am Research Professor at the Radiation and Nuclear Safety Authority, Helsinki, Finland. I am also Guangbiao Professor at the Zhejiang University School of Medicine in Hangzhou, China and Adjunct Professor of Biochemistry at the University of Helsinki in Finland. I have received M.Sc. (1978) and D.Sc. (1983) form the Jagiellonian University in Krakow, Poland and Ph.D. (1990) from the Helsinki University in Finland. I spent two sabbatical periods in the USA: 1990-91 as a Visiting Scientist at Georgetown University in Washington, DC and 1997-99 as a Visiting Assistant Professor at Harvard University in Boston, MA. I study the biological effects of mobile phone radiation using high-throughput screening techniques (HTST) of proteomics to identify the radiation-affected proteins. In 2005 I organized and conveyed the first ever scientific session on the topic of "Genomics, Transcriptomics and Proteomics" at the

* This is an edited, reformatted and augmented version of testimony given on September 14, 2009 before the Senate Committee on Appropriations, Subcommittee on Labor, Health and Human Services, Education and Related Agencies.

bioelectromagnetics meeting held in Rhodes, Greece. Since then, I have lectured on this topic around the world (Europe, USA, India, Japan, China, South Korea, Australia, and South Africa). I have co-organized and co-chaired two World Health Organization's (WHO) workshops and several other conferences around the world. I was an invited expert in the revision of the WHO Research Agenda on mobile phones in 2003 and 2006. I am a member of the Steering Committee of the Swiss National Science Foundation programme on non-ionizing radiation and the Management Committee of the European Cooperation in the Field of Scientific and Technical Research Action BM0704 "Emerging EM F (Electromagnetic Fields) Technologies and Health Risk Management". Within this Action I am chairing the Task Group on use of high-throughput screening techniques in mobile phone research. I was the Co-Chair of the Technical Program Committee for the joint meeting of the Bioelectromagnetics Society (BEMS) and European BioElectromagnetics Association (EBEA) - BioEM 2009, Davos, Switzerland, which is the largest gathering of bioelectromagnetics scientists. I am also the Co-Chair of the Technical Program Committee for the forthcoming 2010 Annual Meeting of BEMS in Seoul, South Korea. In 2006 - 2009 I was the member of the Board of Directors of the Bioelectromagnetics Society and in 2009 I received Leadership Award from the Bioelectromagnetics Society and the European BioElectromagnetics Association. I am an Associate Editor of the journal Bioelectromagnetics. I am reviewer of research grants in the field of bioelectromagnetics for various organizations in the United Kingdom, Holland, Switzerland, Austria, China and South Africa. I co-authored over 90 publications. I comment on current issues in bioelectromagnetics in my Science Blog: http://betweenrockandhardplace.wordpress.com/.

MY RESEARCH IN THE FIELD OF BIOLOGICAL EFFECTS OF MOBILE PHONE RADIATION

I and my research group have worked in the field of biological and health effects of mobile phone radiation for the past 10 years. The main focus of our research has been and is determining whether mobile phone radiation, at levels permitted by the current safety standards, induces biological effects. Proving or disproving this is of paramount importance because if this radiation can not induce biological effects then it can not cause any health effects either. On the other hand, the induction of biological effects does not automatically mean that these will cause a health hazard. Our research consists of both, laboratory experiments and as experiments on human volunteers.

The early sign of a living cell reaction to an external stress factor (e.g. chemicals, radiation, heat, etc.) is activation of stress response designed to protect cells from the damage to its vital functions. Activation of stress response by mobile phone radiation suggests that mobile phone radiation is recognized by the cells as a stress factor that might interfere with normal cellular physiology. Our published research suggests that mobile phone radiation might activate cellular stress response in human endothelial cells, cells that are lining blood vessels [1,2]. Though, our yet unpublished research suggests that the activation of stress response depends much on the conditions of the experiment. Based on our observations we presented hypothesis that the activation of stress response in endothelial cells lining blood vessels in brain might impair function of blood-brain barrier. This barrier, consisting among others of endothelial cells, selectively regulates passage of molecules from blood stream to the brain. Impairment of function of blood-brain barrier might allow passage of such molecules, present in blood stream, that might be damaging for the brain cells. The issue of mobile phone radiation effect on blood-brain barrier is still unresolved, with several animal studies suggesting existence of an effect and several animal studies suggesting the lack of it.

The other part of my group's research is that we suggested that using high-throughput screening techniques of transcriptomics[1] and proteomics[2] could be used to identify potential molecular targets of mobile phone radiation. This so called "discovery science" approach seems to be particularly suited for elucidation of potential mobile phone radiation-associated health hazard issue because it might reveal effects that are not possible to predict based on the present knowledge of the biological effects of mobile phone radiation. Using high-throughput screening techniques of transcriptomics and proteomics we have shown that mobile phone radiation might alter gene expression and expression and activity of proteins. The number of affected genes and proteins appeared to be small and the confirmation of the impact of these changes on human physiology requires further studies [3,4,5]. The changes in gene and protein expression appeared to be dependent on the genetic makeup of the cells. Furthermore, changes in protein expression that were induced by mobile phone radiation appeared to be induced not only in human cells grown in laboratory but also in the skin of human volunteers, as preliminarily suggested by our pilot study [6].

In summary, the basic finding of my group's research is that it appears that mobile phone radiation induces biological response in human cells and these responses might alter cell physiology. However, these findings do not yet prove that there exists health hazard. Further laboratory studies and studies using human volunteers are necessary to clarify this issue.

RELIABILITY OF THE SCIENTIFIC EVIDENCE

There is continuously ongoing controversy whether the users of mobile phones should be concerned about the health safety of the radiation (radiofrequency-modulated electromagnetic fields) emitted by these devices [7,8,9]. The International Committee of Electromagnetic Safety (ICES) of the Institute of Electrical and Electronics Engineers (IEEE), the International Commission on Non-Ionizing Radiation Protection (ICNIRP) and the World Health Organization (WHO) Electromagnetic Fields (EM F) Project are assuring users that there is no proven health risk and that the current safety limits, on the radiation emitted by mobile phones, protect us all [10,11,12]. However, based on the available scientific evidence, the situation might be not that clear.

When evaluating the possible health effects of mobile phone radiation, as with any other environmental factor, no matter naturally occurring or man-made, there are needed several types of scientific evidence such as (i) the possible mechanism how the effect is induced in living organism, (ii) *in vitro* laboratory studies that confirm the existence of the biophysical and biochemical mechanism of the effect, (iii) animal studies, including long term effects and toxicology, (iv) human volunteer studies, and (v) epidemiological evidence of the effect on human population at large.

Each type of the evidence is of different significance and value for the estimation and proof of human health effect. It is considered that the most important evidence is provided by epidemiology, followed by human volunteer studies and animal experiments. *In vitro* laboratory evidence does not directly inform about the possible health impact of any tested factor but it provides information about the possible cellular-level mechanism of the effect and it helps to understand whether the hypothesized biophysical mechanism of the effect is plausible. Knowing the mechanism of the effect increases reliability of human, animal as well as *in vitro* evidence. In the ideal situation, all five types of evidence point into the same direction. This is, however, not always the case. For example electromagnetic fields emitted by the power-lines were classified by International Agency for Research on Cancer (IARC) as a possible carcinogen (category 2B) based only on the epidemiology and there is no clearly supporting evidence for such effects from the animal and *in vitro* studies [13].

In case of the electromagnetic radiation emitted by the mobile phones the scientific evidence is contradictory. In each area of investigation (epidemiology, human volunteers, animal studies, laboratory *in vitro* experiments and biophysical mechanisms) there are both positive and negative studies and by the sheer numbers, the negative studies outweigh the positive ones. This is commonly referred as the "weight of the evidence" that is pointing out to the no-effect-conclusion because the outcome of the majority of published research studies is negative. This argument is used, and often abused, to support the notion that there are no proven effects on health below the present safety standard limits [10,11,12].

So far, the vast majority of the research examining the biological effects of mobile phone radiation has focused on the possibility of induction of genotoxic effects, cancer or impairments in embryonic development. At the same time, the discussion continues on whether mobile phone radiation could cause effects that, although not able to develop into life-threatening disease, could become detrimental to the quality of life. These non-life-threatening effects might include such ailments as e.g. sleep disorders, headaches or allergy-like symptoms. Furthermore, the vast majority of the presently conducted epidemiological, animal, and *in vitro* studies focus on the possibility of induction of cancer. Therefore, independently of their outcome, these studies will provide information only about the cancer and will be unable to neither give mobile phone radiation "a clean bill of health" or to reliably show whether mobile phone radiation exposure can be associated with any non-cancer health risk.

EPIDEMIOLOGICAL EVIDENCE

Epidemiological studies are considered as the most important in evaluation of human health risk. However, due to their low sensitivity in detecting health effects within the population, epidemiology is unlikely to be able to conclusively determine whether weak stimulus, such as mobile phone radiation, causes cancer. There are numerous biases involved in estimation of the health risk by epidemiology, such as: selection bias, misclassification bias, recall bias, and the effect of the developing disease on mobile phone use. Furthermore, there are methodological considerations in epidemiological studies that are unsolved at the moment, such as: no evidence based exposure metric, low duration of mobile phone use, and no evidence-based selection of end-points for epidemiological studies [14].

Further complication with the epidemiological evidence, used to examine the possibility of the increased incidence in brain cancer within the population due to mobile phone radiation exposure, is the long latency period (over 10 years) between the induction and possible diagnosis of brain cancer. Therefore, not surprisingly, majority of the to date executed epidemiological studies, covering at the longest the period of the first 10-years after the start of use of mobile phone, can not be expected to show link between brain cancer and mobile phone radiation, even if it would exist, because of the length of tumor latency period.

Great hopes for finding out the answers whether there is a link between brain cancer and mobile phone radiation exposure were with the European Union 5th Framework Programme-funded "INTERPHONE Project" [15]. This project is a multinational case-control study that was set-up to investigate whether mobile phone radiation increases the risk of cancer and whether it is carcinogenic. The study focused on tumors development in brain and head tissues that are the most exposed to mobile phone radiation and can generate glioma, meningioma, acoustic neurinoma and partoid gland tumors. The study was conducted in 13 countries: Australia, Canada, Denmark, Finland, France, Germany, Israel, Italy, Japan, New Zealand, Norway, Sweden and the United Kingdom (divided in South and North). The hallmark of this project was supposed to be the common core protocol that would allow to pool the data obtained in all 13 countries and to perform one statistical analysis that would consist of sufficient numbers of cases and controls to be meaningful and informative [15]. However, it appears there were numerous exceptions from the common core protocol that put in doubt the reliability and to some extent the validity of the pooled-analysis.

Furthermore, the definition of the ***regular user*** is another major problem of the INTERPHONE protocol. A person who makes at least one phone call per week for the period of at least 6 months is defined as a ***regular user***. This means that a person who makes 24-25 calls over 6 months period (one call per week) is put into the same category with a ***regular user*** who makes 24-25 calls per day - as e.g. many business people do. This means that the category of ***regular users*** consists of those who are only very little exposed to mobile phone radiation but also of those who are very heavily exposed. It means that, if we assume that the heavy exposure to mobile phone radiation could cause any health effects, then these effects will be down-graded by analyzing in the same group of very low- and very high-exposed study subjects. Therefore, the

INTERPHONE's definition of "***regular user***" mixes "***heavy users***" with "***occasional users***". Therefore, considering the definition of "***regular user***" that might dilute the effect, it is somewhat surprising that there are INTERPHONE studies suggesting the increase in brain cancer among the long-term "***regular users***" (>1o years) [16,17]. These results are still statistically uncertain and might either be showing a real link or might be just a statistical "glitch" due to small sample numbers. However, these results were obtained using definition of "***regular user***" as specified above - meaning that any effects induced by mobile phone radiation were, most likely, substantially "diluted" by the presence of very low exposed users among the so-called "***regular users***".

The results of the pooled analysis of the INTERPHONE results are still unpublished due to lack of consensus in interpretation of the results of all 13 national projects. The release of the final results is already delayed for several years (since 2006) due to the disagreements over the conclusions among the different research groups. ***Summa summarum***, it appears that we have to wait for many years for the more reliable answers from epidemiologists. Therefore, the presently available epidemiological evidence has very limited usefulness for setting of human safety standards and for convincingly justifying them.

In my opinion epidemiological evidence is not sufficiently reliable to conclude that human health either is or is not at risk. I think that at this time any statement suggesting that there "is a health risk" or that there "is no health risk", based on the epidemiological evidence, is premature and not reliably supported by the available scientific evidence. However, even though the data are of insufficient quality to draw reliable conclusions, the existence of data suggesting the possibility of increased risk of brain cancer is, in my opinion, sufficient to advice precaution and to request further research to clarify this issue.

The epidemiological case-control studies so far have failed to provide reliable evidence. The new, currently ongoing European cohort study (COSMOS project) is again expected to provide more reliable answers than those of the INTERPHONE project. However, some scientists doubt it because we are dealing with a very weak stimulus and epidemiological approaches might have too low sensitivity to reliably detect small changes in the vast population of mobile phone users. Epidemiological studies are certainly needed but we should in parallel focus on human volunteer studies, using different sub-populations of mobile phone users. Such studies could determine whether mobile phone radiation has any significant effect on normal human physiology. This area has not been studied sufficiently so far.

NON-EPIDEMIOLOGICAL HUMAN STUDIES

The human volunteer studies have focused on mobile phone radiation effects on e.g. cognition, blood pressure, headaches, skin allergy-like symptoms, sleep disorders or direct recognition, by the exposed subject, whether mobile phone is on or off [18]. In these studies the volunteers are not aware when they are exposed to mobile phone radiation and when not. One major set back of these studies, however, is that the experimental conditions and the exposure and measurement equipment may psychologically affect behavior of the volunteers during the experiments and the obtained information might become subjective and unreliable [18]. There is only a single study which provides objective information without asking the study subjects how/what they feel during the exposure. In this study effects of mobile phone radiation on human skin were examined at the molecular level [6]. The outcome of this study suggests that it might be possible that mobile phone radiation alters expression of some proteins in human skin. This single available study has obvious limitations as well as advantages. The main limitation is that, because of the shortage of funding, only pilot study using 10 volunteers was possible to execute. The most obvious advantage of this study is that it has used proteomics approach - simultaneous screening of the expression of hundreds of proteins and, this way, it has shown that mobile phone radiation might affect physiology of human tissue in vivo.

It is certain that just such human volunteer studies, using methods of proteomics, transcriptomics and other biochemical analyses, are urgently needed to demonstrate whether human body (tissues, organs) responds, or not, to mobile phone radiation. Such studies will provide information as to which proteins and genes react to mobile phone radiation exposure. With this information it will be possible to formulate new, knowledge-based, hypotheses for health risk studies in humans [2,19].

To this time, we do not have available objective information whether human body recognizes mobile phone radiation (at levels permitted by the current safety standards) as an external stressor and responds to it at molecular level. Such responses are prerequisite for any physiological/health-related responses. Therefore, because of the lack of studies that would provide unbiased information whether the human body responds to mobile phone radiation, it is problematic to consider that the presently available scientific evidence and based on it current safety standards protect all users of mobile phones.

The effects of mobile phone radiation on children and recommendations of prudent use of mobile phones by children are one of the hottest topics in the discussion of the safety of mobile phones [7,8,9,20]. According to ICNIRP and WHO EM F-Project, the present safety limits protect children: "...in the opinion of ICNIRP, there is neither need nor any justification for a specific approach to the protection of children or other special groups of the population." [21]. However, the scientific basis of such assuring statements seems insufficient. There are no published studies where the effects of mobile phone radiation on development or health of children would have been examined, as revealed by search in the specialized EM F-Portal database (www.emf-portal.de). The scientific basis for such statement comes from the studies examining newborn or young animals. Such data, obtained with young mice or rats at low dose mobile phone radiation, is not sufficient to reliably assure that human children will not be affected.

In the spring of 2008, the European Union 7th Framework Programme for funding research had a specific call for research projects that would examine effects of mobile phone radiation on children. However, results of this research (epidemiological project MOBIKIDS) will not be ready for several years to come.

The presently used safety standards may very well protect the majority of mobile phone users. However, there likely exists a sub-population of people with different sensitivity to mobile phone radiation (not to confuse with the self-diagnosed so-called Electromagnetic Hyper Sensitivity - EHS). It is known that due to genetic variability among people, the same stimuli (medication, radiation, chemicals, allergens, environmental factors) may elicit responses of differing severity in different people [22]. Finding out such sensitive subpopulation and defining it might be possible using the proteomics and transcriptomics [19]. However, such studies are missing and, to my knowledge, no such studies are ongoing or planned or have received funding.

Animal Studies

Another large group of studies used to determine human health risk are animal experiments. Especially the toxicology studies are of interest for establishing of human health risk and for setting human safety standards. In ***sensu stricto*** toxicology studies, animals are exposed to a large overdose of tested chemical, which would not be normally encountered by human being in real situation, and examined for any detrimental effects.

However, such toxicology studies are not possible to perform for mobile phone radiation (microwaves). Large overdose of this "agent" will heat animal and in extreme cases might simply "cook it". Thus, the so far performed and, so-called by the authors, "toxicology" studies are not exactly what they are claimed to be. These are studies where animals were exposed for different periods of time to levels of mobile phone radiation similar to those that are permitted by the human safety standards, and that do not cause heating of the animal. Thus, the usefulness of such studies to estimate human health risk, as compared with the ***sensu stricto*** toxicology is limited. What remains are studies where animals are long-term exposed to similar doses of mobile phone radiation as humans using mobile phones. However, direct extrapolating results of such long-term animal studies, performed at low doses of mobile phone radiation, to human health risk is almost impossible. It is known that although humans and animals possess many of the same genes, the functions of the same genes might differ and some of the same cancer types will be regulated by different genes in animals and in humans. This causes that some of the cancers that will appear in animal will not appear in humans and vice-versa [23]. Thus, the animal "toxicology" studies seem to be of limited use for safety standards setting because they are not ***sensu stricto*** toxicology.

LABORATORY *IN VITRO* STUDIES

The majority of research on the biological effects of mobile phone radiation has been done in laboratory ***in vitro*** studies and the vast majority of the conducted research has focused on cancer. Some of the ***in vitro*** studies suggest, although do not prove, that mobile phone radiation might alter cell physiology e.g. by triggering cellular stress response, causing DNA damage, altering gene and protein expression [7,8,9]. However, there are also studies that do not indicate such effects [7,8,9]. One of the reasons for such discrepancy might be, as postulated in some studies, that the transcriptome- and proteome-dependency of the cell response to mobile phone radiation. It means that genes and proteins expressed (=active) at the time of exposure to mobile phone radiation might play a decisive role in the induction of the effects [4,5]. Laboratory ***in vitro*** studies are of great value for discovering the biochemical mechanism of the effect. They provide support for human and animal studies, but can not be directly used in setting of human health safety standard.

MECHANISMS: THERMAL VS . NON-THERMAL EFFECTS OF MOBILE PHONE RADIATION

In 1998, ICNIRP published the "Guidelines for limiting exposure to time-varying electric, magnetic and electromagnetic fields (up to 3oo GHz)." [11]. In these guidelines, any biological effects of mobile phone radiation induced by temperature rise of up to 1 oC were considered as harmless and this approach became the basis for setting the safety standard that is still in force. At the same time the research field of non-thermal (athermal) effects was described as confusing, as stated in the ICNIRP guidelines: "*...**Overall, the literature on athermal effects of AM electromagnetic fields is so complex, the validity of reported effects so poorly established, and the relevance of the effects to human health is so uncertain, that it is impossible to use this body of information as a basis for setting limits on human exposure to these fields**.*"

According to the committees setting safety standard (ICES & ICNIRP) such small temperature increases of 0.1 - 0.3 oC induced by the mobile phone radiation exposure are meaningless when considering human health risk. The fact that the human body's temperature is in the morning ca. 1 oC lower than in the evening is commonly presented as a supportive piece of evidence. This is seen to suggest that the increase of the body temperature by up to 1 oC can not be hazardous to health because it is within the normal physiological range. However, entirely another question is if small temperature increases induced by mobile phone radiation could be meaningful, from the point of view of the potential induction of biological effects? And yes, the temperature increases are small but do we know enough about their nature and kinetics to be convinced that they are unimportant for the physiology of e.g. brain? I think that the issue might not be so simple. There are two "problems" to consider.

Firstly, the temperature of human body indeed rises over the course of the day by up to 1 oC and the process is harmless. However, this process can not be directly compared with the local temperature increases caused by the mobile phone radiation. Such comparisons are omitting the time-scale of the event, its location and the involved physiological factors. To increase the temperature of the whole body it takes time (even hours). The process is preceded and accompanied by the production of humoral mediators that inform tissues and organs that the temperature increase is happening and cells have time to prepare protective responses. However, in case of the mobile phone radiation, the microwaves act locally by penetrating deeply into the body tissue (e.g. brain) and the temperature increase happens "instantaneously" and without any humoral mediators' warning. The exposed brain cells are suddenly, without any warning, warmed up. This is a non-physiological event for the brain cells to which these cells have not been prepared throughout the evolutionary development. Thus, the direct comparison of the slow and uniform increase in body temperature by classical heating (e.g. sauna, hot-tube or sun-bathing) or by physiological processes (e.g. daily temperature fluctuations, physical exercises or fever) with mobile phone radiation-induced rapid and localized increase in temperature might not be justified.

Secondly, the radiation emitted by mobile phones might induce temperature hot spots within the exposed biological material, i.e. small areas where temperature might rise more than in the neighboring areas. We can detect and measure hot spots on the macro-scale but we do not have yet technology to measure whether hot spots are created on the micro-scale (sub-cellular-scale). Presently, dosimetry and modeling of the distribution and intensity of mobile phone radiation in the brain uses as a model plastic container molded in the form of half-head and filled with "physiological solution" consisting of water, salt and sugar. Such model represents human head with skull (plastic mold) and brain (water solution of salt and sugar). However, it is a great oversimplification of the reality.

Living tissues and cells are not homogenous environments but they are compartmentalized into cells and sub-cellular size volumes (organelles) that are delineated by lipid-containing hydrophobic membranes. Charged biological molecules and ions, unlike in the above mentioned "head model", are not distributed within the tissue or cell uniformly and can't travel freely. Thanks to the membranes and their selective transport mechanisms the distribution of molecules and ions in cells is non-uniform and produces electric gradients that play a paramount role in physiological functioning. Strong electromagnetic fields can disrupt the function of selective transport mechanisms of the membranes and cause profound physiological changes (e.g. electroporation).

The question to be answered is what happens when such hydrophobically compartmentalized environment is exposed to weak electromagnetic stimulus like mobile phone radiation. Will such exposure lead to formation of thermal hot spots on sub-cellular scale, because the free flow of charged molecules and ions is prevented by the selective transport mechanisms? Formation of such sub-cellular hot spots could cause changes in certain functional areas of the cell that eventually could lead to alterations in cell physiology. There have been identified the so-called temperature-sensing molecules in the cell membranes. When activated by temperature change these molecules send stress signals that may reach cell nucleus and may affect expression of genes and, as a consequence, alter cell physiology. However, we still do not know whether the activation of cellular stress response, observed in some studies, is caused by the activation of temperature sensing molecules in cell membrane by the mobile phone radiation or some another event.

In my opinion, it is possible to expect that the mobile phone radiation might affect cells by the combination of thermal and non-thermal (if they exist) mechanisms. Thermal effects, induced by mobile phone radiation, should not be automatically regarded as unimportant in context of health risk evaluation because their occurrence and kinetics are different from the harmless physiological warming up of the body. Unfortunately, the presently available technologies do not yet permit to measure temperature or mobile phone radiation distribution on sub-cellular scale. On the macro-scale of groups of thousands of cells, that are presently measurable, such sub-cellular hot spots would not be detectable. From the point of view of the safety of mobile phone users, the issue of induction of biological (and possibly health) effects by thermal or non-thermal mechanisms should be put to rest. Continuous talk of the harmlessness of thermal effects induced by temperature rise up to 1 oC and the continuous dispute whether non-thermal effects exist is misdirecting science and is taking attention from the important issue - if there are any effects induced by mobile phone radiation at levels permitted by the present safety standards that can alter normal physiology.

NEGATIVE VS . POSITIVE STUDIES AND THE "WEIGHT OF EVIDENCE" ISSUE

It is not only my personal observation that the negative studies seem to be accepted as such, without too much scrutiny, whereas the positive studies are examined in every detail to determine why the result is positive. Hence, the positive studies are not treated equally with the negative ones, even though also the negative studies might include erroneous results or interpretations. Moreover, only the positive studies are demanded to be replicated before they can be accepted as valid scientific evidence. This replication requirement is of course the correct approach, but it should be applied, at least to some degree, also to negative studies. At least the negative studies that are considered as providing the crucial evidence of no-effect should be replicated. An error in study design, execution, data analysis or interpretation might lead not only to positive but also to negative result. Furthermore, many of the positive studies are not even being attempted to be replicated and of course negative studies are not replicated at all. However, if the replication of the positive study is attempted then, commonly, the protocol of the replication study has so many modifications, introduced to improve the quality. That causes that the outcome of such "replication" study is difficult, if not impossible, to compare with the original one. As often happens, the outcome of the so-called replication study differs from that of the original study. However, the failed replication might be either because of incorrect (unreliable) result of the original study or because of the modifications introduced in replication study. Usually, this question remains unanswered but the final result is often claimed to be: in summary, the original study has not been replicated (= is not valid evidence).

Another important issue that comes up when analyzing positive and negative studies is the "weight of evidence". To me this term is often abused by those who wish to disregard scientific studies showing that mobile phone radiation can induce biological effects. We continuously hear that there were done thousands of studies on mobile phone radiation. However, this number is grossly exaggerated because it refers to research at all microwave-frequencies. For example the applicability of the results obtained using radiation frequency of microwave ovens might not necessarily be directly applicable to the mobile phone-emitted microwaves. There is still ongoing discussion whether it is possible to transpose results of experiments done with one frequency of microwaves to other frequencies. To me, in order to be relevant, the studies should be performed using actually mobile-phone-emitted microwaves. The number of such studies, which were done using mobile phone-emitted microwaves, is available from the EM F-Portal database (http://www.emf-portal.de/) that is maintained by the Research Center for Bioelectromagnetic Interaction at the University Hospital of the Aachen University in Germany. This specialized database lists as of September 8th, 2009, total of 631 studies that explicitly investigated the biological and health effects of mobile phone-related microwave frequencies. Therefore, in my opinion, the number of the executed studies is not sufficiently large to create reliable basis for any conclusive statements about the existence or the absence of the health risk associated with the use of mobile phone. These 631 studies include studies that do not show any biological effects of mobile phone radiation but also studies that show induction of such effects. However, because the majority of the published studies show no effect, it is commonly suggested that this "weight of evidence" supports the notion that there are no biological effects and no health risk. In this context, the newly designed, and about to start in the USA, large animal study is considered by some scientists as unlikely to have impact on science concerning mobile phone effects because of the "weight of evidence" provided by the earlier published studies. In short it means that, even well designed, well executed state-of-the-art study with best available radiation exposure dosimetry, might not be sufficient to cause any change in thinking about mobile phone radiation effects. The reason is that the earlier published studies, of which many were poorly designed or executed or had poor dosimetry design, provide "weight of evidence" against any effects. Single well done study is not enough to provide sufficient proof but also a large number of scientifically weaker studies should not be outweighing it.

Conclusions on the Reliability of Scientific Evidence

In the present situation of the scientific uncertainty, the statements assuring that there are not proven health effects and, because of it, the use of mobile phones is safe are premature. In my opinion the current safety standards are not reliable in the context of the lack of studies on human volunteers, children and on effects of long-term exposures in humans. The purpose is not to discourage people from using the mobile phone technology. However, we should remember that we have still remaining large gaps in the knowledge of the mobile phone radiation effects on humans. We urgently need well designed comprehensive molecular level human volunteer studies to close these gaps in the knowledge. In the meantime, it is wise to support the use of precautionary measures in every day dealings with mobile phones in order to, whenever reasonably possible, limit the body exposure to mobile phone radiation.

Advisories for Mobile Phone Users in Finland

The Finnish Radiation and Nuclear Safety Authority (STUK) is a regulatory authority, research centre and expert organisation whose mission is to protect people, society, the environment and future generations from the harmful effects of radiation (www.stuk.fi). STUK belongs to the administration of the Ministry of Social Affairs and Health in Finland. STUK has issued two advisories for mobile phone users. The first one in 2004, as part of the Nordic countries advisory, and the second one in 2009 as STUK's own advisory focused on children using mobile phones. Both advisories mention the uncertainty of the scientific evidence that causes the need for precaution in the use of mobile phones. Excerpt from the Common Nordic Countries View:

> "The existing knowledge gaps and the prevailing scientific uncertainty justify a certain precautionary attitude regarding the use of handsets for mobile telephony. Due to the widespread use of mobile phones even a very small risk could have consequences for public health. Because of the lack of knowledge in certain fields of research the Nordic authorities find it is wise to use, for instance, a hands-free kit that reduces the exposure to the head significantly. This information should be addressed both to adults, young people and children. It is important that parents inform young people and children about how to reduce the exposure from mobile phones."

Excerpts from the STUK advisory "Children's mobile phone use should be limited":

> "In relation to mobile phones children are given special consideration. They will have more time to use a mobile phone for a longer period of time than adults who started using mobile phones more than 10 years ago. The long-term risks from the use of mobile phones can not be assessed before the phones have been in use for several decades. Additionally, children's brains are developing up to the age of 20 years."
>
> "...Children's mobile phone use could be, restricted in the following ways:
>
> - favoring the use of text messages rather than calls,
> - parents limiting the number of calls and their duration,
> - children can be advised in the use of hands-free devices, which reduces the exposure significantly. When communicating on the hands-free device the phone should be kept a couple of centimetres away from the body,
> - talking in an area with low connectivity or in a moving car or a train should be avoided.
>
> However STUK does not deem it justified to ban children's use of mobile phones altogether. As mobile phones also promote security, since it facilitates easy communication with parents.

If an adult is concerned about their own exposure, it can be reduced in the same way as mentioned above for children."

THE FUTURE RESEARCH NEEDS

We need a few well-designed studies, executed by consortia of scientists, not by single research groups. These studies should be aimed at proving or disproving whether human body responds to mobile phone radiation and whether the response is of a sufficient magnitude to alter normal human physiology. In spite of years of research into human health risk of mobile phone radiation, we still do not have the answer to the fundamental question: whether human bodies (tissues, organs) react to mobile phone-emitted microwaves. If human bodies do not react to mobile phone radiation on molecular level, then there will never be any health problem. But, at present, we are still missing science to prove it. We need studies where human volunteers will be exposed to mobile phone radiation and, thereafter, examined for changes on molecular level looking at gene and protein expression and activity changes using methods of transcriptomics and proteomics. Furthermore, we need to look if there are any changes in internal organ physiology by e.g. sampling various body fluids.

The other area that needs urgent attention is the search for mechanisms of the observed biological effects. In particular, the sub-cellular level mobile phone radiation dosimetry needs to be developed to answer the question whether absorption of energy by sub-cellular structures could be a trigger mechanism for some biological effects.

Such studies should be well funded so that scientists will not need to make short-cuts in science because of the lack of funds to perform all of the needed experiments and in a sufficient number of replicates. This is often the case now and that is why so many of inadequate quality and non-informative studies have been published; and provide the distortion to the "weight of evidence". However, getting funds is a problem. Continuous assurances from ICES, ICNIRP and WHO, that thousands of studies have been done and that mobile phone radiation does not cause any known health risk and the safety standards protect us all, have caused that the funding agencies are not interested to sponsor new projects. However, these assurances are weakened by the lack of sufficient scientific evidence to support them.

REFERENCES

[1] Leszczynski D, et al. Differentiation 70, 2002, 120-129
[2] Leszczynski D, et al. Proteomics 4, 2004, 426-431
[3] Nylund R, & Leszczynski D. Proteomics, 4, 2004, 1359-1365
[4] Redmondini D, et al. Proteomics 6, 2006, 4745-4754
[5] Nylund R, & Leszczynski D. Proteomics 6, 2006, 4769-4780
[6] Karinen A, et al. BMC Genomics 9, 2008, 77
[7] Jokela K, et al. Report STUK-A161, 1999
[8] Stewart Report Mobile Phones and Health. 2000. (http://www.iegmp.org.uk)
[9] SSI Independent Expert Group on Electromagnetic Fields. 2008. SSI Reports 2008; 12: 1-70
[10] IEEE Std C95.2-1999 IEEE Standard for Radio-Frequency Energy http://standards.ieee.org/reading/ieee/std_public/description/emc/C95.2-1999_desc.html
[11] ICNIRP. 1998. Health Physics 1998; 74: 494-522
[12] WHO; http://www.who.int/peh-emf/standards/EMF_standards_framework%5b1%5d.pdf
[13] IARC. IARC Monogr Eval Carcinog Risk Hum 2002; vol. 80
[14] Kundi M. Env. Health Perspectives 2009; 117: 316-324
[15] Cardis E, et al. Eur J Epidemiol 2007; 22: 647-664
[16] Schilz J, et al. Am J Epidemiol 2006; 163: 512-520
[17] Lahkola A, et al. Int J Cancer 2007; 120: 1769-1775
[18] Roosli M. Environ. Res. 2008; 107: 277-287
[19] Leszczynski D, & Meltz ML. Proteomics 2006; 6: 4674-4677
[20] Schilz J. Bioelectromagnetics 2005; Suppl 7: S45-50
[21] Vecchia P. Bioelectromagnetics 2005; Suppl. 7: S157-160
[22] Scincariello F, & De Rosa CT. Eur J Oncol 2007; 12: 155-170
[23] Hamad NM, et al. Genes & Dev. 2002; 16: 2045-2057

End Notes

[1] Transcriptomics examines in broad sense expression and activity of genes

[2] Proteomics examines in broad sense expression, structure and activity of proteins

In: Cell Phone Use and Health Risks
Editors: C. Kempson and E. Rahm
ISBN: 978-1-62257-948-8

Chapter 4

TESTIMONY OF SIEGAL SADETZKI, M.D., DIRECTOR, CANCER AND RADIATION EPIDEMIOLOGY UNIT, GERTNER INSTITUTE, ISRAEL. HEARING ON "HEALTH EFFECTS OF CELL PHONE USE"*

PREFACE

I am greatly honored to participate in such an important forum at the United States Senate, and appreciate the invitation to present my research and the efforts made by the Israeli Ministry of Health to address the issue of cell phones and health. In this written statement, I will offer my opinion on a subject which affects a substantial and growing proportion of the world's population, and which I believe has great significance for public health.

I currently serve as the Head of the Clinical Epidemiology Department at the Sheba Medical Center and Director of the Cancer & Radiation Epidemiology Unit at the Gertner Institute, Israel, which are affiliated to the Sackler School of Medicine at Tel-Aviv University (where I hold an appointment as a senior lecturer). I also serve as the Principal National Investigator for Tinea Capitis (ringworm) Research. These studies are conducted in abidance with a law that calls for evaluating health outcomes of

---

* This is an edited, reformatted and augmented version of testimony given on September 14, 2009 before the Senate Committee on Appropriations, Subcommittee on Labor, Health and Human Services, Education and Related Agencies.

treatment with ionizing radiation given for Tinea Capitis in the 1950s to about 50,000 children (most of them immigrants who came to Israel in the 1950s). The law was established in 1994, to compensate the irradiated population for late radiation outcomes. In accordance with this law, my responsibilities include advising the Ministry of Health regarding compensation for diseases which have been found to be causally related to the irradiation, and developing medical guidelines for follow-up for this unique population.

I am thus actively involved in research and in advising the Chief Director of the Ministry of Health for determining health policy concerning ionizing and non-ionizing radiation and cancer.

I am an official member of several national professional committees established to advise the Ministry of Health: these include among others the National Council for Prevention, Diagnosis and Treatment of Cancer, the National Council for Diagnostic Imaging and the Advisory Committee on Cancer Epidemiology for the Director General. In addition, I have been nominated to participate in several ad hoc committees dealing with specific issues such as guidelines for Pap smear tests in Israel etc.

In my role as advisor, I authored *the Israeli Ministry of Health guidelines for the use of cellular phones in adults and children,* as well as the Director General's Statement concerning the Tinea Capitis Compensation Law (1994), and participated in the drafting of the Director General's Statements concerning guidelines for imaging procedures using ionizing radiation in children, and risks of radiation in cardiac imaging procedures.

Since 2005, I have been a member of the Brain Tumor Epidemiology Consortium (BTEC), an open scientific international forum organized to promote studies on brain tumors, for which I served as the European president between the years 2007-2009.

I have conducted research on both ionizing and non-ionizing radiation in children and adults with colleagues throughout the world, and am currently involved in a number of collaborative studies of brain and other cancers funded by the NIH, USAMRAA (United States Army Medical Research Acquisition Activity), the European Community and other grant institutions. About 85 peer reviewed articles have been published from these studies.

STUDIES ON CELL PHONE USE & CANCER

For more than a decade, I have been involved in studies of the possible association between the use of cell phones and the risk of malignant and benign brain tumors, tumors of the acoustic (hearing) nerve and tumors of the salivary gland. I served as the principal investigator of the Israeli part of the International collaborative "INTERPHONE" study, the largest epidemiological study conducted to date on this topic.

This study (conducted between the years 2000-2005) was coordinated in the IARC (International Agency for Research on Cancer) with the participation of 13 countries (Australia, Canada, Denmark, Finland, France, Germany, Israel, Italy, Japan, New Zealand, Norway, Sweden and the UK). Several manuscripts on the methodology of this study have been published [1-9]. While the overall results of the total group have not yet been published, several reports from individual centers, as well as a pooled analysis of 5 centers, have been published, each presenting their individual results [10-17].

I am currently leading the Israeli part of another multi-national collaborative effort to investigate the risk of brain cancer from exposure to radiofrequency fields in childhood and adolescence (MOBIKIDS). This study was funded following the European Union's call for proposals within the Seventh Framework Programme (ENV.2008.1.2.1.1.). The study, with the participation of 16 centers from several countries in Europe, Israel, Canada, Australia and New Zealand, will be conducted during the years 2009-2014. The study population will comprise indi viduals aged 10-24 years, among whom about 1,900 will be cases diagnosed with malignant and benign brain tumors and about 3,900, healthy controls.

It should be noted that in addition to cancer, there are other medical outcomes that also should be explored, including influence on brain acti vity, beha vioral changes, learning patterns, emotional well-being, immunologic pathways, fertility etc. Cultural, social, and other non-medical outcomes should also be assessed.

The Israeli INTERPHONE Study — Country Specific Results

In February 2008, the Israeli results of the assessment of the association between cell phone use and risk of benign and malignant parotid gland tumors (the major sali vary gland located at the anterior border of the external ear, 4-10 mm deep in the skin surface, in the area close to where phones are often held) were published in the American Journal of Epidemiology. This nationwide Israeli study followed the core protocol of the INTERPHONE study and was extended to include a larger study population. While, as reported in other studies, no increased risk of these tumors was seen for the *total* group, *consistent elevated risks were shown in complementary analyses restricted to conditions that may yield higher levels of exposure.* An elevated risk of salivary gland tumors was seen among people who used cell phones for more than 10 years, especially when the phone was usually held on the same side of the head where the tumor was found, and when use was relatively heavy [16].

Our findings of a higher risk with greater exposure (as expressed by laterality of use, more frequent use, and longer duration of use), are consistent with basic public health research criteria for what is referred to as a dose response relationship—the greater the dose or use of cell phone in our study (or exposure), the greater the response (i.e., the risk of developing a tumor). The Israeli population is characterized by exceptionally high le vels of cell phone use as expressed by the cumulati ve number and duration of calls. This situation created an important opportunity for studying the effects of relati vely high patterns of use, which did not exist in most other populations.

Methodological Limitations of Cell Phone Studies

Our findings are in line with several other studies that demonstrated an increase in risk of developing malignant brain tumors and acoustic neurinoma (tumors on the nerve that controls hearing) associated with relatively long-term use (>10 years), and with cell phone use on the same side of the head as the tumor [12,17-21].

However, *there is a debate in the scientific community about the interpretation of these findings and whether these observations reflect a true association or stem from the numerous methodological problems that accompany epidemiological studies in general, and those on cell phones specifically.*

Epidemiological studies have the advantage of investigating human beings in real life situations. However, considering the complexity of exposures and health conditions in the population, epidemiological studies suffer from a handful of methodological challenges that need to be resolved in order to ensure valid results.

The challenges of investigating a relatively new research topic like cell phones, and a devastating disease such as brain tumor, which can affect cognition and recall, are complex. Many articles and reports have criticized the methodology of the published studies, including that of the INTERPHONE [2, 3, 7-9, 22, 23]. Unfortunately, I can not cover all of these methodological issues in this testimony; nevertheless, I would like to briefly mention some of them.

Duration of at least 10 years is the minimal time needed for solid cancer studies and, in the case of brain tumors it may reach 30-40 years. To illustrate this point, I would like to use the proven association between ionizing radiation and brain tumors that was assessed in the atomic bomb studies as an example. The A-bomb was dropped on Hiroshima and Nagasaki in August 1945. While an excess risk of leukemia among survivors was observed in the 1950's, and an increased risk of solid tumors was detected in the 1960's, no elevation in risk of brain tumors was noted among survivors for many years. The first report demonstrating brain tumors among the survivors was only published in 1994 (almost 50 years later) and the first quantitative data for all intracranial tumors, in 2002 [24, 25]. Consequently, even in the mid 1970's, it was not evident to the scientific community that ionizing radiation can cause brain tumors. Since widespread cell phone use really began only in the mid 1990's, the follow up period currently available in most published studies is only a little more than 10 years, *which is insufficient to detect an effect, if one exists.* Moreover, in most studies, the "dose" of the exposure, as expressed by cumulative number of calls and duration of calls was also rather limited during the initial years of use. *As current use is characterized by greater levels of exposure* it increases the chance of finding an effect.

As exposure to Radio-Frequency from cell phones is localized (98% of the energy is absorbed in the brain hemisphere on the side where the phone is used), analysis of cell phone use in relation to location of tumor is necessary for the interpretation of results of these studies.

Valid tools for exposure measurements and assessment are crucial for accuracy of results. However, many difficulties exist in the exposure assessments of cell phones which have to account for laterality of use, period of use, patterns of use, technological aspects of the phones and networks, other exposures to electromagnetic radiation, not to mention the constantly changing technology. Lack of precision in these measurements could lead to errors and biases in the results. The data needed for these studies is usually taken from questionnaires based on self-reporting which are subject to inaccuracies due to problems such as recall. The alternati ve use of billing records is also limited due to technical problems in abstracting historical records, shared SIM cards, discrepancies between owner and user of the phone, as well as the inability to determine side of use and use of hands-free de vices which dramatically lowers the exposure. Novel exposure metrics developed by a multi-disciplinary expert team including epidemiologists, physicists, and industry workers who are familiar with this technology are needed to facilitate accurate measurement of this complicated exposure.

FUTURE SCIENTIFIC NEEDS

Despite these difficulties, it is of utmost importance to see that such studies continue, gi ven the broad range of uses and exposures that exist around the world today. I believe I am correct in saying that there is a consensus that the information currently a vailable is insufficient. Therefore, additional extensi ve research is necessary to clarify this issue. As the United States of America has always been a leader in conducting scientific and medical research, your choice of this topic as a high priority is needed in order to make significant progress in this field. In view of the many challenges that complicate research on cell phones, I am con vinced that a multidisciplinary multinational effort conducted in various populations, and benefiting from the experience acquired in pre vious studies is the optimal road to clarifying the health risks.

It is important to note that the study populations of the research carried out so far were limited to adults. While at the time when cell phone use began, only adults used cell phones, since the beginning of the 21st century, increasing numbers of children have become users and even owners of cell phones. *This population requires special attention since children have been found to be more susceptible to developing cancer following exposures to known carcinogens.* Furthermore, the brain of a child is not just a smaller version of that of an adult and the radiation absorption in their head is different than adults. The observation of greater susceptibility at younger ages has been consistently shown in numerous studies and for a variety of known carcinogens. In our research as well, on the effects of ionizing radiation used to treat children with tinea capitis, the risk for malignant brain tumors 40 years after treatment was in versely related to age at time of irradiation. Children irradiated under 5 years of age exhibited a risk that was 4.5 times higher than that of children who had not undergone irradiation, while those irradiated at ages 10-15 had a risk that was 1.5 times higher than the non exposed [26].

PUBLIC HEALTH CONSIDERATIONS

There are now 4 billion people, including children, using cell phone technology. Consequently, e ven if there is only a small indi vidual risk per person, the great number of users, together with the increasing amount of use, could eventually result in considerable damage. *Therefore, until definitive answers are available, some public health measures with special emphasis for children should be instituted.* Preventive steps implemented for other technologies associated with risks, such as dri ving, provide a good example. We all use cars but, in order to reduce the risk of accidents, legislation has been passed concerning the use of seat belts, speed limits, minimum age requirements for driving licenses, and car tests.

Based on the findings of my work and on reports from scientists in other nations, the Israeli Ministry of Health issued a statement on cell phone use that adopted the *precautionary principle* (that briefly means: "better safe than sorry") [27]. This approach rests on the important public health concept that: In case of doubt regarding the data, it is far better to prevent harm *using simple and low cost measures* than to wait for long-term results that confirm a health hazard that has already occurred. Therefore, in such cases, we must be prepared to act before scientific certainty has been achie ved (using reasonable and low cost acti vities).

The recommendations of the Israeli Ministry of Health [28], include several simple, low cost measures that should be taken to minimize exposure, such as using speakers and earphones, hands-free devices when dri ving, and reducing the use of cell phones in areas where reception is weak. Special attention was gi ven to children, who have been found to be more susceptible to developing cancer following exposures to known carcinogens.

Guidelines have also been published in other countries. I will not mention all of the existing guidelines but will gi ve some examples with special emphasis on recommendations for children. It is interesting to mention that the definition of a child varies from country to country. The French Health Ministry has published a warning against excessive use of cell phones, especially among children, and has recommended a voiding calls when reception is weak and while driving, as well as for holding the phone away from sensitive areas of the body by using speakers or hands-free devices. It is also considering the possibility of banning the sale of cell phones designed for children under the age of 6, prohibiting the ad vertising of mobile phones directed at children less than 12 years of age, and requiring the manufactures to develop cell phone that allow only sending and receiving messages [29, 30].

The Finnish Radiation and Nuclear Safety Authority (STUK) has recently suggested restricting the exposure of children to cell phones by encouraging the use of text messages rather than calls and the use of hands-free devices through which the phone is kept away from the body. Additional recommendations include ad vising parents to limit the number and duration of calls made by their children, and a voiding calls in areas with low reception or in a moving car or train. It is important to note that STUK does not believe that banning cell phone use in children is justified, as cell phones also promote security since they facilitate easy communication with parents [31].

The Toronto Public Health Department suggested that parents should think twice before giving their children (especially pre-adolescents) cell phones. It also recommended that landlines should be used, whenever possible, while cell phones should be used only when absolutely essential. It was also suggested that the length of cell phone calls be limited, and that headsets or hands-free options be used whenever possible [32]. In India, the Health Minister has recently suggested that people should not talk on a cell phone continuously for more than one hour a day, and that hands-free technology could reduce the side effects of excessive use [33].

Recommendations have also been published in other countries such as the UK, Russia, Germany, and Belgium [34]. The issue of what constitutes appropriate policy in this regard is not, strictly speaking, a scientific matter, but one of judgment. Scientists fulfill their role by providing concrete, independent information on potential hazards, while those charged with policy development have the more difficult job of recommending what to do about the problem, as science continues to evolve.

Summary

Advances in technology have improved the quality of our li ves in many ways and these changes have been especially dramatic in the area of communication. I believe that cell phone technology, which has many advantages, and can save lives in emergency situations, is here to stay. However, the question that needs to be answered is not if we should use cell phones but how we should use them? It is my hope that the issues raised in this forum have enabled the distinguished legislators in this hall to appreciate the need to promote research that will increase our understanding of the potential adverse effects and take actions which will ensure the safe and responsible use of cell phones, while research and technology continue to evolve. Thank you again for inviting me and for bringing attention to this important issue.

REFERENCES

[1] Cardis E. Kilkenny M. for the International Study Group, International Agency for Research on Cancer. International case-control study of adult brain, head and neck tumours: results of the feasibility study. *Radiat Prot Dosim* 1999; 83: 179-183.

[2] Vrijheid M, Cardis E, Armstrong BK, Au vinen A, Berg G, Blaasaas KG, Brown J, Carroll M, Chetrit A, Christensedn HC, Deltour I, Feychting M, Giles GG, Hepworth SJ, Hours M, Ia varone I, Johansson C, Klaeboe L, Kurttio P, Lagorio S, Lonn S, McKinney PA, Montestrucq L, Parslow RC, Richardson L, Sadetzki S, Salmiinen T, Schuz J, Tynes T, Woodward A. Validation of short-term recall of mobile phone use for the interphone study. *Occup En viron Med* 2006; 63: 237-243.

[3] Vrijheid M, Deltour I, Krewski D, Sanchez M, Cardis E. The effects of recall errors and of selection bias in epidemiologic studies of mobile phone use and cancer risk. *J Expo Sci En viron Epidemiol* 2006; 16: 371-84.

[4] Cardis E, Richardson L, Deltour I, Armstrong B, Feychting M, Johanssen C, Kilkenny M, McKinney P, Modan B, Sadetzki S, Schuz J, Swerdlow A, Vrijheid M, Au vinen A, Berg G, Blettner M, Bowman J, Brown J, Chetrit A, Collatz CH, Cook A, Hepworth S, Giles G, Hours M, Ia varone I, JarusHakak A, Klaeboe L, Krewski D, Lagorio S, L6nn S, Mann S, McBride M, Muir K, Nadon L, Parent M-E, Pearce N, Salminen T, Schoemaker M, Schlehofer B, Siemiatycki J, Taki M, Takebayashi T, Tynes T, van Tongeren M, Vecchia P, Wiart J, Woodward A, Yamaguchi N. The INTERPHONE study: design, epidemiological methods, and description of the study population. *Eur J Epidemiol* 2007; 22: 647-664.

[5] Cardis E, Deltour I, Mann S, Moissonnier M, Taki M, Varsier N, Wake K, Wiart J. Distribution of RF energy emitted by mobile phones in anatomical structures of the brain. *Phys Med Biol* 2008; 53: 2771-2783.

[6] Wake K, Varsier N, Watanabe S, Taki M, Wiart J, Mann S, Deltour I, Cardis E. The estimation of 3D SAR distributions in the human head from mobile phone compliance testing data for epidemiological studies. *Phys Med Biol* 2009; 54: 5695-5706.

[7] Vrijheid M, Armstrong BK, Bedard D, Brown J, Deltour I, Ia varone I, Krewski D, Lagorio S, Moore S, Richardson L, Giles GG, McBride M, Parent ME, Siemiatycki J, Cardis E. Recall bias in the assessment of exposure to mobile phones. *J Expo Sci En viron Epidemiol* 2009; 19:369- 81.

[8] Vrijheid M, Richardson L, Armstrong BK, Au vinen A, Berg G, Carroll M, Chetrit A, Deltour I, Feychting M, Giles GG, Hours M, Ia varone I, Lagorio S, Lönn S, McBride M, Parent M-E, Sadetzki S, Salminen T, Sanchez M, Schlehofer B, SchLiz 3, Siemiatycki 3, Tynes T, Woodward A, Yamaguchi N, Cardis E. Quantifying the impact of selection bias caused by non-participation in a case-control study of mobile phone use. *Ann Epidemiol* 2009; 19: 33-42.

[9] Vrijheid M, Mann S, Vecchia P, Wiart 3, Taki M, Ardoino L, Armstrong BK, Au vinen A, Bédard D, Berg-Beckhoff G, Blaasaas KG, Brown 3, Chetrit A, Collatz-Christensen H, Combalot E, Cook A, Deltour I, Feychting M, Giles GG, Hepworth S, Hours M, Ia varone I, 3ohansen C, Krewski D, Kurttio P, Lagorio S, Lönn S, McBride M, Montestrucq L, Parslow RC, Sadetzki S, SchLiz 3, Tynes T, Woodward A, Cardis E. Determinants of mobile phone output power in a multinational study implications for exposure assessment. *Occup En viron Med* 2009 (Ahead of print - 21.5.09).

[10] Lönn S, Ahlbom A, Hall P, Feychting M; Swedish Interphone Study Group. Long-term mobile phone use and brain tumor risk. *Am J Epidemiol* 2005; 161: 526—535.

[11] Christensen HC, SchLiz 3, Kosteljanetz M, Poulsen HS, Boice 3D 3r, McLaughlin 3K, 3ohansen C. Cellular telephones and risk for brain tumors: a population-based, incident case-control study. *Neurology* 2005; 64: 1189-95.

[12] Schoemaker M3, Swerdlow A3, Ahlbom A, Au vinen A, Blaasaas KG, Cardis E, Christensen HC, Feychting M, Hepworth S3, 3ohansen C, Klaeboe L, Lönn S, McKinney PA, Muir K, Raitanen 3, Salminen T, Thomsen 3, Tynes T. Mobile phone use and risk of acoustic neuroma: results of the Interphone case-control study in fi ve North European countries. *Br J Cancer* 2005; 93: 842-8

[13] SchLiz 3, Böhler E, Berg G, Schlehofer B, Hettinger I, Schlaefer K, Wahrendorf 3, Kunna-Grass K, Blettner M. Cellular phones, cordless phones, and the risks of glioma and meningioma (Interphone Study Group, Germany). *Am J Epidemiol* 2006; 163: 512—520.

[14] Hours M, Bernard M, Montestrucq L, Arslan M, Bergeret A, Deltour I, Cardis E. Téléphone mobile, risque de tumeurs cérébrales et du nerf vestibuloacoustique: l'étude cas-témoins INTERPHONE en France (Cell Phones and Risk of brain and acoustic nerve tumours: the French INTERPHONE case—control study; Rev Epidemiol Sante Publique), Revue d'Epidémiologie et de Santé Publique 2007;55:321-32.

[15] Schlehofer B, Schlaefer K, Blettner M, Berg G, Böhler E, Hettinger I, Kunna-Grass K, Wahrendorf 3, SchLiz 3; Interphone Study Group. En vironmental risk factors for sporadic acoustic neuroma (Interphone Study Group, Germany). *Eur J Cancer* 2007; 43: 1741-7.

[16] Sadetzki S, Chetrit A, Jarus-Hakak A, Cardis E, Deutch Y, Du vdevani S, Zultan A, Novikov I, Freedman L, Wolf M. Cellular phone use and risk of benign and malignant parotid gland tumors - a nationwide case-control study. *Am J Epidemiol* 2008; 167: 457-467.

[17] Lahkola A, Salminen T, Raitanen J, Heinavaara S, Schoemaker MJ, Christensen HC, Feychting M, Johansen C, Klaeboe L, Lönn S, Swerdlow AJ, Tynes T, Au vinen A. Meningioma and mobile phone use—a collaborati ve case–control study in fi ve North European countries. *Int J Epidemiol* 2008 August 2 (Epub ahead of print).

[18] Hardell L, Carlberg M, Soderq vistF, Mild KH. Meta-analysis of long-term mobile phone use and association with brain tumours. *Int J Oncol* 2008; 32: 1097-103.

[19] Kan P, Simonsen SE, Lyon JL, Kestle JRW. Cellular phone use and brain tumor: a meta analysis. *J Neurooncol* 2008; 86: 71-78.

[20] Lahkola A, Tokola K, Au vinen A. Meta-analysis of mobile phone use and intracranial tumors. *Scand J Work En viron Health* 2006; 32: 171-177.

[21] Lahkola A, Au vinen A, Raitanen J, Schoemaker MJ, Christensen HC, Feychting M, Johansen C, Klaeboe L, Lonn S, Swerdlow AJ, Tynes T, Salminen T. mobile phone use and risk of glioma in 5 North European countries. *Int J Cancer* 2007; 120: 1769-1775

[22] Ahlbom A, Feychting M, Green A, Kheifets, Sa vitz DA, Swerdlow AJ and ICNIRP. Epidemiologic evidence on mobile phones tumor risk – a review. *Epidemiology* 2009: 20: 639-652.

[23] Morgan LL, Barris E, Newton J, O'Connor E, Philips A, Philips G, Rees C, Stein B, Eds. Cellphones and Brain Tumors 15 Reasons for Concern Science, Spin and the Truth Behind Interphone August 25, 2009. online publication: http://www.radiationresearch.org/pdfs/reasons$_{v}$us.pdf

[24] Shibata S, Sadamori N, Mine M, Sekine I. Intracranial meningiomas among Nagasaki atomic bomb survi vors. *Lancet* 1994; 344: 1770.

[25] Preston DL, Ron E, Yonehara S, Kobuke T, Fujii H, Kishikawa M, Tokunaga M, Tokuoka S, Mabuchi K. Tumors of the nervous system and pituitary gland associated with atomic bomb radiation exposure. *J Natl Cancer Inst* 2002; 94: 1555-1563.

[26] Sadetzki S, Chetrit A, Freedman L, Stovall M, Modan B, Novikov I. Longterm follow-up for brain tumor development after childhood exposure to ionizing radiation for tinea capitis. *Radiat Res* 2005; 163: 424-432.

[27] Commission of the European Communities. Communication from the Commission on the precautionary principle. February 2, 2000 (http://ec.europa.eu/dgs/health consumer/library/pub/pub07 en.pdf , accessed September 6, 2009).

[28] Israeli Ministry of Health. Recommendations of the Ministry of Health regarding cell phone use. August 11, 2008 (http://www.health.go v.il/pages/default.asp?PageId=4408&catId=838& maincat=46 , accessed September 6, 2009).

[29] Radiation fear on mobile phone usage- Warns French government. (http://www.mgovworld.org/News/radiation-fear-on-mobile-phone-usage-warns-french-go vernment , accessed September 10, 2009).

[30] French government bans ad vertising of mobiles to children. New limits will be placed on radiation levels amid fears of increased risk of cancer from phone use. Geoffrey Lean, En vironment Editor. January 11, 2009 (http://sn.im/k14u1 , accessed September 6, 2009).

[31] Radiation and Nuclear Safety Authority: Children's mobile phone use should be limited. January 7, 2009 (http://www.stuk.fi/stuk/tiedotteet/en GB/news 527 , accessed September 6, 2009).

[32] Toronto Public Health urges parents to limit children's cellphone use. July 15, 2007 (http://www.healthyontario.com/NewsItem Details.aspx? newsitem id=12 01 , accessed September 6, 2009).

[33] Azad warns against excessive use of cell phones. July 23, 2009 (http://timesofindia.indiatimes.com/news/India/Azad-warns-against-excessi ve-use-of-cell-phones/articleshow/4808807.cms, accessed September 6, 2009).

[34] En vironmental Working Group. Cell Phone Radiation Science Review, Section 3: Government Action on Cell Phone Radiation Levels. (http://www.ewg.org/cellphoneradiation/Government-Action, accessed September 10, 2009).

In: Cell Phone Use and Health Risks
Editors: C. Kempson and E. Rahm
ISBN: 978-1-62257-948-8

Chapter 5

STATEMENT OF LINDA S. ERDREICH, PH.D, SENIOR MANAGING SCIENTIST, EXPONENT HEALTH SCIENCES. HEARING ON "HEALTH EFFECTS OF CELL PHONE USE"*

INTRODUCTION

I am a Senior Managing Scientist in Exponent's Health Sciences Center for Epidemiology, Biostatistics, and Computational Biology, and I have 30 years of experience in environmental epidemiology and health risk assessment. I earned a Ph.D., in Epidemiology from the University of 0klahoma in 1979, and an M.S., Biostatistics and Epidemiology in 1977. My work includes assessing epidemiological research and integrating this information with that from other disciplines for qualitative and quantitative risk assessments. I have prepared analyses of complex epidemiological evidence for environmental and occupational chemicals, radiofrequency energy, electric and magnetic fields (EMF), and stray voltage, and have published in the peer reviewed scientific literature. I have been particularly active in updating standards regarding non-ionizing radiation, both low frequencies (EMF) and radio frequencies. I have served on advisory committees to government, regulatory organizations, and industry regarding health risk assessments of chemicals and electromagnetic fields.

* This is an edited, reformatted and augmented version of testimony given on September 14, 2009 before the Senate Committee on Appropriations, Subcommittee on Labor, Health and Human Services, Education and Related Agencies.

BACKGROUND

Mobile phones operate using radio waves. Radio waves, or radiofrequency (RF) energy, is a range of the electromagnetic spectrum that includes AM and FM broadcast radio, television, and many other devices and technologies including cordless phones, baby monitors, radar, and microwave ovens. Visible light is also part of the electromagnetic spectrum, but is at a higher frequency and shorter wavelength than RF. RF energy is not "radiation" in the same sense as used for high frequency X rays, because the energy of RF is so much lower and is unable to change the DNA of cells. Although RF energy is sometimes referred to as "EMF" the contemporary usage of EMF refers primarily to the electric and magnetic fields associated with electricity from power lines and all electric devices. Electricity operates in the extremely low frequency (ELF) range, 60 cycles per second (60 Hz), in the United States. To avoid confusion, I will use RF in my discussion of mobile phones.

Standard Scientific Methods Are Used to Assess Possible Risks to Human Health

The standard scientific approach used to determine whether an exposure source, such as to RF energy, poses a health risk, is to look at all of the available research, including both studies that have reported effects, and those that did not. The goal is an objective, comprehensive review, in which the strengths and weaknesses of each study are evaluated, and more weight is given to studies of better quality. This approach is designed to ensure that reviewers do not single out studies, consciously or inadvertently, to support a preconceived opinion. Then, all of the studies are evaluated together to arrive at a conclusion. This is the method that I have used for evaluating the RF research and for other assessments throughout my career.

The relevant research to be considered includes a broad spectrum of scientific research that uses different approaches to study potential effects of RF energy on humans. These different approaches have different strengths and limitations and provide complementary information: laboratory studies in cells and in animals, experimental studies of human volunteers, and epide miologic studies of human populations. For this reason, scientific organizations convene panels of independent experts from the various areas of expertise (e.g. health physics, engineering, toxicology, clinical medicine, and epidemiology) relevant to the topic. Many scientific organizations consider pertinent studies to be those reports of scientific research or reviews that have been published or accepted for publication in the peer-reviewed scientific literature (IARC, 2006; WH0, 2007).

Independent Scientific Organizations Worldwide Have Reviewed the Research

Independent scientific organizations worldwide have reviewed the research and proposed exposure limits. Many studies have been conducted over the past 50 years to examine whether exposure to RF energy has adverse effects on health, and to determine allowable levels of exposure. Several scientific organizations have reviewed the laboratory and epidemiologic research to assess the potential for health effects from RF exposure, and to set exposure limits to ensure occupational and public safety. These expert groups have included scientists with diverse skills to reflect the different research expertise required to answer questions about RF energy and health. Numerous government agencies and professional organizations have reviewed the science related to potential health effects from using wireless phones. While the specific conclusions vary, all of the reports that assess the evidence using multidisciplinary panels and a comprehensive approach reach similar conclusions; the current scientific evidence does not demonstrate that wireless phones cause cancer or other adverse health effects.

The Federal Communications Commission (FCC) and the Food and Drug Administration (FDA), the agencies with regulatory authority over radiofrequency emissions in the U.S., have both concluded that the current scientific evidence does not indicate that there are health hazards from using a wireless phone. The FCC's website states that "[t]here is no scientific evidence that proves that wireless phone usage can lead to cancer or a variety of other problems, including headaches, dizziness or memory loss."(www.fcc.gov/ cgb/cellular.html#evidence) The FDA's website similarly states that "[t]he weight of scientific evidence has not linked cell phones with any health problems."(http://www.fda.gov/Radiation-EmittingProducts/RadiationEmitting ProductsandProcedures/HomeBusinessandEntertainment/CellPhones/ucm 1162 82.htm)

In September 2008, the National Cancer Institute (NCI), the U.S. government's principal agency for cancer research, published a Fact Sheet on Cellular Telephone Use and Cancer Risk that concluded that there is no consistent link between cellular telephone use and cancer. The NCI also stated that "incidence data from the Surveillance, Epidemiology, and End Results (SEER) program of the National Cancer Institute have shown no increase between 1987 and 2005 in the age-adjusted incidence of brain or other nervous system cancers despite the dramatic increase in use of cellular telephones.[2] "http://www.cancer.gov/cancertopics/factsheet/Risk/cellphones.

The conclusions of these U.S. agencies are similar to the conclusions reached in reports prepared by various commissions and agencies around the world, including for example:

- The Australian Radiation Protection and Nuclear Safety Agency (2009)

"There is essentially no evidence that microwave exposure from mobile telephones causes cancer, and no clear evidence that such exposure accelerates the growth of an already-existing cancer." http://www.arpansa.gov.au/mobile phones/index.cfm

- Health Canada (2007)

"There is no convincing scientific evidence that RF exposures have any link to cancer initiation or promotion. The body of peer-reviewed literature in this area overwhelmingly demonstrates a lack of linkage, and where the few reports of linkage effects were found, some may be attributed to factors other than RF energy."

- The Health Council of the Netherlands (2008)

"The Committee maintained its conclusion that no causal link has thus far been demonstrated between health problems and exposure to electromagnetic fields generated by mobile phones or base stations for mobile telephony." http://www.gezondheidsraad.nl/sites/default/files/200902.pdf (English version starts at page 65).

- The Scientific Committee on Emerging and Newly Identified Health Risks of the European Commission (2009)

"Overall, research indicates that mobile phone use does not increase the risk of cancer, especially when used for less than ten years." http://ec.europa.eu/health/opinions2/en/electromagnetic-fields/index.htm#3

- The World Health Organization (2006)

"Considering the very low exposure levels and research results collected to date, there is no convincing scientific evidence that the weak RF signals from base stations and wireless networks cause adverse health effects." http://www.who.int/mediacentre/factsheets/fs304/en/index.html

The United Kingdom's Health Protection Agency and New Zealand Ministry of Health's National Radiation Laboratory also have reached similar conclusions after reviewing the available science.

In September 2009, the International Commission on Non-Ionizing Radiation Protection's (ICNRP) Standing Committee on Epidemiology published a scientific review of all of the available epidemiologic evidence on wireless phones and brain tumors. That review concludes:

> "In the last few years, the epidemiologic evidence on mobile phone use and risk of brain and other tumors of the head has grown considerably. In our opinion, overall the studies published to date do not demonstrate a raised risk within approximately 10 years of use for any tumor of the brain or any other head tumor." (Ahlbom et al, 2009)

Conclusion

Based on my review of the epidemiologic studies and consideration of experimental data in animals, I agree with the conclusions of the scientific organizations: The current scientific evidence does not de monstrate that wireless phones cause cancer or other adverse health effects.

References

Ahlbom Anders; Feychting Maria; Green Adele; Kheifets Leeka; Savitz David A; Swerdlow Anthony J Epide miologic evidence on mobile phones and tumor risk: a review. Epidemiology (Cambridge, Mass.) 2009;20(5):639-52.

Health Council of the Netherlands (HCN). BioInitiative Report. U-5601/EvR/iv/673-L1 Publication nr 2008/17E. September 2008.

International Commission on Non-Ionizing Radiation Protection (ICNIRP). Exposure to high frequency electromagnetic fields, biological effects and health consequences (100 kHz-300 GHz), 2009.

World Health 0rganization (WH0). Fact Sheet No. 304 - Base stations and wireless technologies, 2006. http://www.who.int/mediacentre/factsheets/fs304/en/index.html

End Notes

[1] *See* http://www.cancer.gov/cancertopics/factsheet/Risk/cellphones.

[2] *Id.*

In: Cell Phone Use and Health Risks
Editors: C. Kempson and E. Rahm
ISBN: 978-1-62257-948-8

Chapter 6

STATEMENT OF DEVRA LEE DAVIS, PH.D., PROFESSOR OF EPIDEMIOLOGY, UNIVERSITY OF PITTSBURGH. HEARING ON "HEALTH EFFECTS OF CELL PHONE USE"*

It is a distinct honor and great privilege to present information before this subcommittee on a matter of tremendous importance to this country and to the world—the relationship of cell phones to our health. This subcommittee is to be commended for bringing public attention to this issue through holding the first Senate hearing on cell phones in about there decades.

My remarks today will draw on my three decades of experience as a public health researcher, teacher and writer. I was the founding director of the Board on Environmental Studies and Toxicology of the U.S. National Research Council, National Academy of Sciences, where I also served as Scholar in Residence, from 1983-1993. I was confirmed by this body as presidential appointee to the U.S. Chemical Safety and Hazard Investigation Board from 1994-1999. More recently I founded and directed the Center for Environmental Oncology at the University of Pittsburgh Cancer Institute from 2 005 until 2 009. I am currently a Professor of Epidemiology in the Graduate School of Public Health. I have served on numerous governmental and international advisory boards, including the National Toxicology Program Board of Scientific Counselors and the World Health Organization, health

* This is an edited, reformatted and augmented version of testimony given on September 14, 2009 before the Senate Committee on Appropriations, Subcommittee on Labor, Health and Human Services, Education and Related Agencies.

indicators program. I have also advised governments in China, Mexico, Brazil, Israel, France, the European Union and local governments throughout the world.

Designated a National Book Award Finalist for my first popular book, When Smoke Ran Like Water, in 2 002, my recent book The Secret History of the War on Cancer was a Newsweek must-read pick for the week and has just been released as a paperback. Both works have been translated into Italian and Chinese and been the subjects of documentary television and public radio and television coverage.

I have held a number of academic appointments including Visiting Professor at Carnegie Mellon University's Heinz School, Honorary Professor, London's School of Hygiene and Tropical Medicine, and expert Advisor to the World Health Organization. I have authored more than 19 0 publications in books and journals ranging from Scientific American to the Journal of the American Medical Association and the Lancet, and the Annals of the New York Academy of Sciences, and have also written for the New York Times, the Los Angeles Times, and other mass media outlets. Recently, I have established the Environmental Health Trust, a nonprofit foundation dedicated to identifying and controlling environmental causes of illness (www.environmentalhealthtrust.org). Among the awards I have received are: The Lisa Zhang Environmental Award of the United Nations 2 008, the Artemis Award from the Euro-American Women's Organization and the Greek Ministry of Foreign Affairs in 2 009, the Women's Leadership Exchange Compass Award, presented by OPEN: The Small Business Network from American Express, for breaking the paradigms of how women are perceived and the first Rachel Carson Award from the Rachel Carson Homestead in 2 008. I was also privileged to be part of the group receiving the Nobel Peace Prize with Al Gore in 2 007 for serving as a Lead Author of the Inter-governmental Panel on Climate Change.

I speak to you today as a scientist, and as a citizen of this great democracy who is also a mother and a grandmother. I am deeply concerned about the absence of a major program of research on cell phones and our health. I want to know, as do the American people, why are other governments acting to warn about the need for safer cell phone use while ours is silent on the matter?

This hearing presents a welcome opportunity to address a subject little discussed in the United States. We know that cell phones have revolutionized our lives for the better. They have radically changed the nature of emergency response and warfare. They have improved our sense of security. But, we must admit that we don't know whether some of their uses place us and our children at risk in the long term and whether there are simple measures to take to reduce those risks. Certainly, we have heard today of growing concerns about the impact on our health from cell phone use from distinguished researchers and national leaders of efforts in Israeli and Finland —countries with great sophistication regarding radar and electronics. We are agreed that additional research is critically needed to clarify the potential hazard of cell phones, which are currently being used by more than half of the world—a great proportion of today's users are under age 3 0.

We have learned that Dariusz Leszcynski, PhD, DSc, who is a Research Professor at STUK — Radiation and Nuclear Safety Authority, Helsinki, Finland and Siegal Sadetzki, MD, MPH Head of the Radiation and Epidemiology Unit at Sheba Medical Center, Gertner Institute of Epidemiology, Tel Aviv Medical University and a key advisor to the Israel Radiation Protection Branch, are conducting important research on the question of what cell phones may mean for our health.

In Finland, Israel, China, and the European Union, governments are officially warning citizens about the safer use of cell phones even while that research is still underway. Similar advisories exist in England, France, and some of the states of India and Russia. We have to ask: What do these countries know that we do not?

There is much that we do not know. It will be expensive to resolve many of the issues that must be addressed. But, we do have some information, as the meeting I am chairing in Washington, D.C. this week makes clear. We know that cell phone radiation, hereinafter referred to as RF, can cause biological impacts in experiments with cell cultures and with laboratory animals at levels that do not produce heat or thermal effects.

We must admit that there is another reason why we in the U.S. lag behind other nations in addressing these problems. As I have documented in my book, The Secret History of the War on Cancer, public discussion in the U.S. about potential cell phone risks remains obscure because of well-honed efforts by some in the cell phone industry to keep us confused.

The question before this body is what is direct evidence at this time on cell phones and health and what do we do while we wait for science to evolve?

That question is not merely a matter that can be answered by science, but will require leadership and a vision of basic public policy. We must ask what is the downside risk of doing nothing to reduce exposures at this point, compared to the risk of acting as other governments have to issue warnings. We must also consider what sorts of policy options should be used to convey information and whether it is appropriate for this government to take specific actions at this time that are in line with those taken by others..

Let me stress that the science on this issue is truly complex. It will be expensive to resolve many of the issues that must be addressed. The battleground has been drawn in the realm of both experimental findings and with regard to public health research. Henry Lai has pointed out that if one examines the funding for studies, a simple pattern emerges—studies funded by industry directly are overwhelmingly negative and find no effect of RF in animals or humans. Those studies that are independently funded and have examined people for a decade or longer tend to be positive and find that RF is linked with a host of ailments, ranging from cardiac disturbances to fatal brain tumors. ,The challenge we face is enormous. While science continues to evolve in its understanding of RF signals and our health what do we do while we wait? That question is not merely a question of science, but of basic public policy.

Science is a complicated discipline. When it comes to evaluating potential hazards in the environment, we do not have the luxury of waiting several decades for scientific clarity. We are forced to take precautionary steps while scientific information becomes clearer. The existence of scientific uncertainty should not become an excuse for inaction. In this regard, the decision to take no action, to do nothing, must be understood as a decision to continue the status quo.

As you are well aware the history of regulatory interests makes it clear that in the matters of lead in gasoline and the control of tobacco had we acted sooner to reduce these hazards millions would have been spared damaged brains and lungs. This history tells us that we are obliged to make good sense of what is known at this time about cell phones and our health.

As this chamber knows full well from the sorry history of tobacco, those who do not like particular scientific findings have proven especially adept at treating science as nothing other than a public relations tool. Yet there is no one reading this testimony who doubts we should have acted sooner to address the hazards of smoking or those of lead in gasoline.

As this body itself determined, the reasons for delays in controlling tobacco had much to do with politics not with science. When President Nixon launched the "War on Cancer "in 1971, he ignored the Surgeon General's 1964 declaration that tobacco was a cause of poor health. And President Jimmy Carter fired Secretary of HEW Joseph Califano in 1979, when Califano had the nerve to declare tobacco public enemy number one. At one point in the 197 0s, the National Cancer Institute was led by a four pack a day smoker who spearheaded an effort to spend taxpayer dollars to develop a safe cigarette. The U.S. government actually spent millions of taxpayer dollars to develop a safe cigarette and continued to subsidize tobacco well into the 199 0s. The government only began to try to restrain this dangerous habit about a decade ago, after the epidemic of smoking-related disease became undeniable and after incontrovertible evidence was revealed of the duplicity of the tobacco industry in manipulating science and regulatory policy on the issue.

As a result of these delays in addressing the dangers of tobacco, the world is now reeling from a massive global epidemic of lung cancer, with more than a million cases expected this year in China alone.

What ***about*** cell phones? What do we really know about their safety? Consider these undisputed facts.

Fact: Brain tumor rates are increasing in young adults in several nations, including this one. Brain cancer deaths are now the leading cause of cancer deaths in children in this nation, Sweden and Australia among others. In truth we cannot attribute this or the puzzling and sad rise in autism to cell phones. But, clearly this is a matter that requires serious attention.

Fact: The Federal Communications Commission sets standards for the amount of radio frequency that can be emitted by a cell phone is based on models of a man's head. And not just your average Joe, but also one who ranked at the top 9 0th percentile of all military personnel in 1988, weighing in at 2 00 pounds, and who held the phone to his ear for six minutes. Also, what few parents know is that radio-frequency signals reach much more deeply into children's thinner and smaller heads than ours-—a fact established through the pioneering work of professor Om P. Gandhi, the leader of the University of Utah's electrical engineering department and confirmed recently by studies developed by Niels Kuster and his colleagues in Austria. While these two engineers may differ on the details of their models of the brain, their work makes it clear that children's brains differ in important ways from those of adults. Their skulls are thinner. Their brains contain more fluid. As a result, even if exposures were identical in the depth of the skull that they reach, the potential for doing damage is much greater with the young brain. Whether the elderly face different risks is also an important question to explore.

Fact: The agency that offers recommendations on cell-phone emissions in the U.S. — the Federal Communications Commission doesn't employ a single health expert. The standards the FCC adopts are based on advice given by outside experts, many of whom work directly for the cell-phone industry. Unlike with drugs that are tested before being used, the Food and Drug

Administration lacks the authority to set standards for cell phones and can only act if a phone is shown to release hazardous signals.

What's wrong with this picture? The award-winning Gandhi worries that all the standards used for phones apply to the "big guy" brain. In 2 004, standards became looser as a result of a new approach that basically doubled the amount of radio frequency that could reach the brain of an adult and quadrupled that reaching a child's. The brain of a child doubles in the first two years of life and keeps on developing until their early 2 0s. Gandhi no longer works with the cell-phone industry and none of his grandchildren, or mine, uses a cell phone.

Fact: Many of the negative studies on cell phones and human health involve short period of exposure with much older phones. Thus, one of the most widely cited studies of cell phone safety is that of the Danish Cancer Society, which studied close to half a million-cell phone users as of the mid 199 0s. They excluded all business users from their study—the group most likely to use the phone the most. They found no increased risk in all the others.

Surely, today's phones and the ways we use them are far different from what went on in Denmark a decade ago, when cell phones were as heavy as small briefcases.

Proving harm in science is not a simple matter, especially when it comes to a technology as powerful and widely used as cell phones are today. The science that is implicating cell phones today includes both experimental modeling like that developed by Gandhi and Kuster, as well as the ever-more perplexing studies of epidemiologists. Coming from two Greek words meaning epi, upon and demos, the people, epidemiologists look for patterns of disease in time and space to make sense of the real world. Studying brain cancer is one of the toughest jobs in epidemiology. What happens to moms and dads where they live and work and what they eat and drink can have an impact on whether children develop brain cancer. So, we know that men or women who work with some pesticides and solvents tend to have children with much higher rates of brain tumors.

But the disease can take forty years to develop in adults. Because most adults with brain cancer don't survive, and those who do are often left with problems of speech and recall—either from the disease itself or from the treatment—we often have to interview their remaining family members about their life histories and try to figure out what could have led to the disease. Few of us really know all the good and bad things we've dealt with in our lives, let alone those of our relatives.

When it comes to sorting through the risks of cell phones, we have been assured by widely publicized reports from what appear to be independent scientific reviewers that there is none. Researchers from the Danish Cancer Society reported in the Journal of the National Cancer Institute in 2 007 that they found no evidence of risk in several hundred thousand persons who had used cell phones. Headlines around the world boasted of this latest finding from an impeccable source published in a first tier scientific journal.

The press coverage of this study tells us a great deal about what journalists and the rest of us who depend so heavily on these phones would like to believe. These headlines appeared within days of publication.

"Cell Phones Don't Cause Brain Cancer" — The Toronto Daily News, December 1 0, 2 006 "Cell Phones Don't Raise Cancer Risk" — Reuters, December 6, 2 006

"Big Study Finds No Link Between Cell Phones, Cancer" — San Jose Mercury News, December 6, 2 006

"Study: Cell Phones Do Not Cause Cancer" — Albuquerque Tribune December 6, 2 006 "Study: Cell Phones Safe" — Newsday December 7, 2 006

"Cell Phones Do Not Cause Cancer" — Techtree.com, India, December 7, 2 006

But let's look at what the Danish researchers actually studied.

They reviewed health records about brain tumors that occurred up to 2 002 of about 421, 000 people who had first signed up for private use of cell phones between 1982 and 1995. A "cell phone user" in the study was defined as anyone who made a single phone call a week for six months during the period 1981 to 1995. In fact, the study first started out with almost 7 00, 000 people, but the researchers kicked out anyone who was part of a business that used cell phones, including those who had used a cell phone for personal purposes for eight years. Think of those early clunker phones with their battery packs, cumbersome cords, and hefty monthly fees—those are the phones first examined in this study. Business users are certain to be those with lots of reasons to some lots of time on the phone. Because researchers could not be sure that only one person used a business phone, they threw out all business phones.

This research design raises a lot of questions. Why did the researchers not look at individual business users—those with far more frequent use of cell phones? Why lump all users together, putting those who might have made a single cell phone call a week with those who used the phones more often? Why stop collecting information on brain tumors that had occurred only as of

2 002? Use of cell phones has grown more than fourfold since that time in many countries, including many parts of the U.S., U.K., France, and Israel.

When you are looking at a large population to find an effect, generally the more people you study, the better your chance of finding something. But if you include lots of people with little exposure along with those with very high exposure, you basically lower your chances of finding any effect at all. Lumping all these various users together is like looking all over a city for a stolen car when you know it's within a five-block radius. Perhaps you'll find what you're looking for, but the chances are greater that you won't.

If you want to find out whether cell phone use causes brain cancer, the higher the use or exposure of those you are examining, the better the odds that you will be able to find whether or not it's made a difference. It's clear that the early analogue phones must be different than the newer digital ones. We hope the difference is big and that those of us using phones today face a lowered risk, but we have no way to know whether this is the case. Some of us believe (and hope) that using speakerphones or earpieces connected by wires—not the hands-free kinds— should reduce our direct exposures, but, again, direct evidence on this is not at hand.

In all circumstances, research works best when we have solid information on the nature of the use or exposure we are looking at. All of us have cell phone bills that provide detailed records of our use, and most of these can be accessed online. These were not used in this Danish study, nor in any study done for or by the industry to date. A gold mine of data lies untapped and so far untappable. Dr. David Servan-Schreiber, a distinguished psychiatrist and medical researcher, and author fo the best-selling book, Anti-Cancer, and I are working with cell phone companies in France and elsewhere to encourage the release of billing records so that epidemiologists can carry out much more sophisticated studies than have been possible thus far..

The Danish study, as the headlines made clear, found no increase in risk of brain cancer for private users of cell phones. The reason the researchers were looking for brain cancer is straightforward. As the authors noted, cell phone signals do penetrate the brain. "During operation, the antenna of a cellular telephone emits radio frequency electromagnetic fields that can penetrate 4-6 cm into the human brain."[i]

There is, however, a vibrant debate over what this absorption into the brain means biologically. We know that the body is electric and that electricity in medicine can be used to heal bones and restart or regularize heart beats. But what about RF signals themselves?

We know that cell phone signals can reach the side of the head where the auditory nerve is located. An earlier Swedish study, compared more than 14 00 people with brain tumors to a similar number without the disease during the time 1997-2 000. They found that tumors of the hearing nerve were three times more frequent in those who had used cell phones for more than a decade.[ii] This difference even passed the demanding scientific test of statistical significance, which essentially shows that the results are not likely to be just some random finding. In 2 004, other Swedish researchers found that long-term users also had significantly more tumors on the hearing nerves than nonusers.[iii] My colleagues at the University of Pittsburgh Graduate School of Public Health and Medical Center have recently confirmed this result, in work that is attached to this testimony—persons using a cell phone for a decade or longer have significantly elevated risks of acoustic neuromas.

Of course, most of the published work in the field is negative, but it turns out that much of this is inconclusive by design. But, is that really the end of the matter? We know that tumors of the hearing nerve and malignant cancers of the brain can take decades to form. All of these widely publicized negative studies have several things in common. One study that was well publicized in 2 000 found no increased risk of brain cancer in cell phone users. There's only one problem with this result-- the average cell phone users in this widely publicized study had a phone for less than three years.[iv] Still, even this limited study found that those who had used phones for this short period of time had twice the risk of a very rare brain tumor—neuroepitheliomatous cancers—the kind that wraps itself around the nerve cells of the lining of the brain—right at the locus that cell signals can reach.

One of biggest challenges to any study of cell phones and human health is that the problems they are trying to understand are inherently complex. Science works best to study one thing at a time, as we do with drugs in clinical trials carefully meting out specific doses and tracking specific responses. But, the world we live in is much messier and more complicated than the elegant one of clinical research. The problems posed by cell phones in the real world are like huge simultaneous equations—mathematical formulas of relationships between multiple unknowns. How can you determine the role of one factor, such as cell phone exposure to the skull, when all others, like diet, workplace conditions, and local air pollution, are changing at the same time and at different rates? How do you take into account the fact that phones themselves have changed design and that peoples' habits in using phones may also change with reports of concerns growing?

Given how broadly cell signals now penetrate our worlds of the coffee shop, traveling discounted buses, airports, and many downtown areas of major cities, where do we find any truly unexposed groups to compare results against? Because cell phone use has grown so fast and technologies change every year, it is as if we are trying to study the car in which we are driving.

Some of the works done in laboratories at the Medical University of Vienna and elsewhere clearly showed that wireless signals could affect the ways cells talk to one another to stay under control—what is called gap-junction communication. Under healthy conditions, cells send messages through proteins and enzymes that keep things in order and tell badly behaving cells to get in line or die. Wireless signals were shown to throw a monkey wrench into this order. Like teenagers, cells that can't communicate well are prone to grow out of control. In essence, the wireless signals promoted a kind of social breakdown among cells.

As Dr. Sadetzki has told us, the human health component of the study of cell phones remains unfinished, and it may well be unfinishable. A major international study of brain cancer in wireless phone users has been underway for nearly a decade, headquartered at the renowned International Agency for Research on Cancer (IARC) of the World Health Organization in Lyon, France. The large study was designed to combine more than three thousand cases of brain tumors from around the industrial world and was supposed to release its results in almost five years ago.

At the core of the IARC project is a major effort to learn from brain cancer patients whether they used cell phones more frequently than did others. The limits of the work are easy to grasp. The ways to overcome them are not. Still, some German findings published in 2 006 are disquieting. Keep in mind that not a single one of these studies is actually using billing records of cell phone use. All of them require people to try to remember their habits over ten years or so.

The German study captured information about the daily lives of people in Mainz, Bielefeld and Heidelberg. What did they have for breakfast regularly? Where did they live? How often did they use the cell phone? For how long? On which ear? These are the sorts of things epidemiologists like me hope you remember. This work contrasted the life experiences and reported cell phone use of 366 people with deadly tumors of the brain called gliomas and 381 with slow-growing, usually benign tumors of the membranes that cover the spinal cord, against some 15 00 people between the ages of 3 0 and 69 who had better luck and did not have brain tumors. When asking both groups about their past and current uses of cell phones, they did not find any increased risk in those who used phones for less than a decade. That was not the end of this work, however, but merely the start.

In this same study, those who reported having used cell phones for ten years or more had twice the risk of coming down with gliomas.[v] This is a tumor that begins in the glial cells of the brain, the non-conducting cells that support the neurons and hold them together. The growth of gliomas can be silent, with symptoms that mimic flu or a headache. But eventually, they become undeniable. People lose speech, sight, movement or hearing, depending on where the tumor starts and where it ends up.

It should be obvious that looking at people with a fatal illness and asking them to try hard to remember what they did up to forty years ago in some cases is not easy. Doing this the day after brain surgery in the hospital is obviously not an ideal situation for obtaining records.

With all the highly automated information governments now assemble to combat terror, including library and cell phone records, what would it take for authorities to allow expert scientists access to privacy-protected, coded computerized records of cell phone use so that we could learn whether our use of cell phones places us at risk from a disease that could be averted through better design and technology?

That's not a question likely to get much attention at this moment, but it is well worth thinking about. The studies to date that have not found a general, clear and consistent risks from cell phones have tended to follow people for short periods of time. Brain tumors can take four decades to become evident. Of necessity, the older studies have for the most part have looked at older technologies over short periods of exposure. With one exception, no researcher has asked about the impact of cell phones on the brains of children and teenagers—one of the fastest growing groups of users in the world today.

Recently, several groups including the Collaborative on EMF Research and the Environmental Working Group and the European Union have issued reports concluding that cell phones cause or greatly increase the risk of brain tumors. What is this based on?

The Interphone study, as Dr. Sadetzki just told us, has not reached a clear conclusion. But every single study that is part of Interphone and has studied people who used phones heavily for a decade has found that where persons have used phones heavily for a decade or longer, there is evidence of a significantly increased risk—literally a doubled risk of malignant brain tumors.

The one researcher to have studied young people who began using cell phones as teenagers, Prof. Lennart Hardell of Sweden, has found that those who started to use cell phones heavily before age 2 0 have 4 to 6 times more brain tumors by the time they reach their 3 0s. This is deeply troubling.

What should we do now?

Representatives of the industry have recently been quoted as saying: "peer-reviewed scientific evidence has overwhelmingly indicated that wireless devices do not pose a public health risk." To me, the absence of definitive evidence on this issue is not proof there is no harm, but a reflection of two things. First, it is hard to conduct epidemiological studies on cell phone users for obvious reasons, as Dr. Sadetzki has explained. Second, there are powerful interests that have kept us from asking and answering important questions. We have also seen repeatedly that the chances a study will label cell phones safe depends chiefly on who pays for the study. (Lai and Roos.)

What does independent research really show? What do the FDA and ACS really say about the matter? On their website, The Center for Devices and Radiological Health (CDRH) of the U.S. Food and Drug Administration (FDA), states: "Available science does not allow us to conclude that mobile phones are absolutely safe, or that they are unsafe. However, the available scientific evidence does not demonstrate any adverse health effects associated with the use of mobile phones."

The FDA and FCC jointly state that those who are concerned should take simple precautions such as using earpieces and speakerphones, while acknowledging that Quote, "The available scientific evidence does not show that any health problems are associated with using wireless phones. There is no proof, however, that wireless phones are absolutely safe." End quote.

But, the FDA site goes on to say the following:

"The scientific evidence does not show a danger to any users of cell phones from RF exposure, including children and teenagers. The steps adults can take to reduce RF exposure apply to children and teenagers as well.

- Reduce the amount of time spent on the cell phone
- Use speaker mode or a headset to place more distance between the head and the ce phone.

Some groups sponsored by other national governments have advised that children be discouraged from using cell phones at all. For example, The Stewart Report from the United Kingdom made such a recommendation in December 2000. In this report a group of independent experts noted that no evidence exists that using a cell phone causes brain tumors or other ill effects. Their recommendation to limit cell phone use by children was strictly precautionary; it was not based on scientific evidence that any health hazard exists."

http://www.fda.gov/Radiation

EmittingProducts/RadiationEmittingProductsandProcedures/HomeBusinessandEntertainment/Ce l lPhones/ucm116331.htm

In 2 001 a review article in the American Cancer Society journal, Cancer written by Howard Frumkin and Michael Thun, a senior epidemiologist with the American Cancer Society in fact took a precautionary approach as well. Noting the absence of clear evidence of harm and the relatively short period of time that studies had been underway, they offered simple advice: "It is impossible to prove that any product or exposure is absolutely safe, especially in the absence of very long-term follow-up. Accordingly, the following summary from the Food and Drug Administration Center for Devices and Radiological Health offers advice to people concerned about their risk:

If there is a risk from these products—and at this point we do not know that there is—it is probably very small. But if people are concerned about avoiding even potential risks, there are simple steps they can take to do so. People who must conduct extended conversations in their cars every day could switch to a type of mobile phone that places more distance between their bodies and the source of the RF, since the exposure level drops off dramatically with distance. For example, they could switch to: a mobile phone in which the antenna is located outside the vehicle, a hand-held phone with a built-in antenna connected to a different antenna mounted on the outside of the car or built into a separate package, or a headset with a remote antenna to a mobile phone carried at the waist. Again the scientific data do not demonstrate that mobile phones are harmful. But if people are concerned about the radiofrequency energy from these products, taking the simple precautions outlined above can reduce any possible risk.[4]

http://www.cancer.cellular$_N$phones$_N$and$_N$risk$_N$of$_N$brain$_N$tumors.asp

In other recent public statements, Thun, who is a Vice President of ACS observed that: "Cellular (cell) phones are a relatively new technology that became widely used in the United States only in the 199 0s. Although they have been studied extensively, we don't yet have information on the potential health effects of very long-term use or usage by children. "

The good news is that manufacturers and beginning to incorporate this advice into information they provide to consumers. The challenge will be to get people to read and act on this information. How many people know that the directions for using Blackberries for the new 4G wireless devices also note that phones should not be kept on the body and that those concerned about children's exposures should take efforts to reduce that exposure. Those with pacemakers are urged to keep the device at least 2 0 centimeters, or about 8 inches away from the chest. Warnings with the new Blackberry smartphones and iPhones state that the phone should be kept .98 inch or 25 millimeters or 5/8 inch or 15 millimeters respectively away from the body and that failure to do so could result in excessive exposure. U.K advisors have urged caution with respect to children.

Recently a number of scientists have confirmed an observation reported by Hungarianj and Australian scientists in 2 004 http://www.getting pregnant.co.uk/use-of-cell-phones-andsperm-count-motility.html and by the Cleveland Clinic last year. http://www.clevelandclinic.org/reproductive researchcenter/docs/agradoc239.pdf

Regularly keeping a cell phone in the pocket produces defects in sperm form and sperm count. Experimental studies have produced similar results, yielding sperm with impeded motility and reduced numbers..

Unfortunately, scientists who have tried to conduct independent research in this area have often found themselves under the gun. Cell phone research became a kind of third rail for many scientists—touch it and you die.

Dr. Om Gandhi for years led studies at the University of Utah evaluating emissions from phones for Motorola and all the major cell phone companies. As a pioneer in modeling of cell phone absorption into the brain, he in 1996 published analyses, which showed that existing models did not protect children's brains. No surprise: he lost all his industry funding. Now, because he's a tenured professor of a certain age, he continues to do the work and has produced further analyses showing he's correct. Recently French Telecom and other industry supported groups have also confirmed what any mother knows--a child's head is much more sensitive thinner, less dense, more fluid, and therefore more vulnerable than that of an adult.

A study by Professors Henry Lai and Singh showed that low levels of RF signals could produce strange defects in DNA in 1994. The industry response? First, they went to the journal where the paper had been accepted and tried to get the paper un-accepted. Then, they hired a PR firm to try to discredit the findings. Then, they gave money to other researchers in an effort to disprove the findings. When this research confirmed their findings that RF could damage DNA, that research was not published. The full story on this can be found at http://www.washington.edu/alumni/columns/march O5/wakeupcall Othtml and is also discussed in my new paperback.

A similar situation is still being resolved in Vienna, where a multi-million dollar multi-laboratory study of the damaging effect of DNA on RF led by

Professor Rudigger and Adlkofer of the Medical University of Vienna was charged with fraud. These charges became headlines around the scientific world. An independent investigation by the university has recently reported that the charges of fraud were not correct. But, the damaging effect of the charges cannot easily be reversed.

I am fortunate that I worked for Ronald B. Herberman at the University of Pittsburgh. In reviewing the evidence provided by the Bioinitiative Report and the European Environment Agency last year, he recognized this issue for what it is—a major chance to promote research while in the meantime preventing harm. The Pittsburgh Advisory and his statement on this issue, along with other background documents can be found at www.environmentalhealth trust.org

As we have heard today, scientific and policy leaders in Israel, France, Finland, Russia and China have since echoed Pittsburgh's precautionary advice.

I would urge the Senate to consider carefully the case for national action. I am encouraged by the fact that the new head of radiological and other devices at the FDA has indicated an interest in examining this issue.

My advice at this point is simple:

As a number of groups have recently urged, Cell phones should have warning labels stating: children's brains need special protection; phones should not be kept on the body, and should only be used with ear pieces or speakerphones. Children should be encouraged to text and not use phones next to their heads.

What about the much needed research? Who should do it and where should the money come from? Obviously, the history of cell phones and public relations tells us we must make this research independent. Creating a cabinet level inter-agency group on cell phone research, like that on climate change, seems a good start.

As to funding, I've got a simple idea that many of my colleagues in industry tell me makes sense. Let's put an extra research fee of $1. 00 on every cell phone for three years and use these funds to support the conduct of a major independent research program to address the questions raised by this panel and by the National Academy of Sciences in its 2 008 report on the subject.

Given the widespread and important role that cell phones play in our lives today and their invaluable use for many purposes, we can and must do better. Our children and grandchildren will thank us if years from now if they are using safer devices because we took the step at this moment in history to create the solid research program to create an improved technology. I am confident that with this hearing a new day of open dialogue has begun and I thank the Senators for making this possible.

End Notes

[i] Joachim Schilz, et al., "Cellular Telephone Use and Cancer Risk: Update of a Nationwide Danish Cohort," *Journal of the National Cancer Institute* 98, no. 23 (2006): 1707-13.

[ii] L. Harde ll, et al., "Case-control study on the use of cellular and cordless phones and the risk for malignant brain tumours," *International Journal of Radiation Biology* 78, no. 10 (2002): 931-936.

iii S. Lonn, A. Ahlbom, P. Hall, and M. Feychting, "Mobile phone use and the risk of acoustic neuroma," *Epidemiology* 15, no. 6 (2004): 653-659.

iv Joshua E. Muscat et al., "Handheld Cellular Telephone Use and Risk of Brain Cancer," *Journal of the American Medical Association* 284, no. 23 (2000): 3001-3007.

v Joachim Schilz et al., "Cellular Phones, Cordless Phones, and the Risks of G lioma and Meningioma (Interphone Study Group, Germany)," *American Journal of Epidemiology* 163, no. 6 (2006): 512-520.

In: Cell Phone Use and Health Risks
Editors: C. Kempson and E. Rahm
ISBN: 978-1-62257-948-8

Chapter 7

TESTIMONY OF OLGA V. NAIDENKO, PH.D., SENIOR SCIENTIST, ENVIRONMENTAL WORKING GROUP. HEARING ON "HEALTH EFFECTS OF CELL PHONE USE"*

Mr. Chairman and distinguished Members of the Subcommittee: My name is Olga Naidenko, and I am a Senior Scientist at Environmental Working Group (EWG), a nonprofit research and advocacy organization based in Washington, DC; Ames, Iowa; and Oakland, California. I thank the members of the subcommittee for holding this important hearing and for the opportunity to testify.

Last week, EWG released the results of a 10-month investigation of more than 200 peer-reviewed studies, government advisories, and industry documents on the safety of cell phone radiation. We found that the studies amassed during the first two decades of cell phone use produced conflicting results and few definitive conclusions on cell phone safety. But the latest research, in which scientists are for the first time able to study people who have used cell phones for many years, suggests the potential for serious safety issues.

Studies published over the past several years find significantly higher risks for brain and salivary gland tumors among people using cell phones for 10 years or longer. The state of the science is provocative and troubling, and more

* This is an edited, reformatted and augmented version of testimony given on September 14, 2009 before the Senate Committee on Appropriations, Subcommittee on Labor, Health and Human Services, Education and Related Agencies.

research is essential. We at Environmental Working Group are still using our cell phones, but we also believe that until scientists know much more about cell phone radiation, it's smart for consumers to buy phones with the lowest emissions.

As of December 2008, U.S. wireless subscribers numbered 270.3 million 99 87 percent of Americans 99 a 30 percent jump in three years (CTIA 2009). Some 60 percent of the global population – four billion people 99 subscribe to wireless services (ITU 2008). As the market for new devices has grown, so has the urgency that cell phone safety be well understood, and that cell phone radiation standards be sufficient to protect public health.

In this testimony we highlight five key areas of concern:

- Consumers have a right to know the level of radiation their phones emit;
- Latest science points to potential risks to children's health;
- Federal standards for cell phone radiation need to be modernized;
- What consumers can do to reduce exposures to cell phone radiation;
- EWG's recommendations to the government, industry, and the public.

1. Consumers Have a Right to Know the Level of Radiation Their Phones Emit

EWG advocates that cell phone companies label their products' radiation output so that consumers can make informed choices at the point of sale, and that the government require this disclosure. Currently, most people are given no information at all about radiation emissions when they purchase a phone.

To fill this information void, EWG's research team created a user-friendly, interactive online guide to cell phone emissions, covering over 1,200 phones currently on the market. Consumers can use this free online database to make informed decisions about which cell phones to buy. The EWG guide uses easy-to-read graphics to illustrate each phone's radiofrequency emissions, enabling consumers to make quick comparisons of radiation output of various wireless devices.

In the 64 hours following the publication of our science review and cell phone radiation database, 442,000 people accessed these materials on our website, collectively viewing 1.4 million online pages. During those same 3 days our findings were reported in 100 news articles and in national and local broadcast news, including the *New York Times*, *NBC Nightly News*, *WebMD*, and *USA Today*. This powerful response from the public and from news media outlets reflects consumers' keen interest in the issue of cell phone safety. Clearly, people are eager to know if cell phones are safe and how they can protect themselves and their families from potential adverse effects of excessive exposure to cell phone radiation.

2. The Latest Science Points to Potential Risks to Children's Health

Prior to 2003, studies of cancer risk and cell phone use produced conflicting results. The Food and Drug Administration (FDA) told consumers that scientists had found no harmful health effects from exposure to cell phone emissions (FDA 2003). But FDA's assurances were based on studies of people who had used cell phones for just 3 years, on average (FDA 2003), not long enough to develop cancer.

At that time, studies had not addressed the risks of longer-term cell phone radiation exposures. The research gap is closing. Recent studies find significantly higher risks for brain and salivary gland tumors among people using cell phones for 10 years or longer. The state of the science is provocative and troubling, especially for the health of children. Among recent findings are the following:

- A joint study by researchers in Denmark, Finland, Norway, Sweden and the United Kingdom found that people who had used cell phones for more than 10 years had a significantly increased risk of developing glioma, a usually malignant brain tumor, on the side of the head they had favored for cell phone conversations (International Agency for Research on Cancer (IARC) 2008; Lahkola 2007).
- French and German scientists reported an increased risk of glioma for long-term cell phone users (Hours 2007; Schuz 2006). Analysis of all published cell phone-brain tumor studies found that people who had used a cell phone for 10 or more years, the overall risk for developing a glioma on the cell phone side of the head increased by 90 percent (Hardell 2009; Kundi 2009).
- Cell phone use for 10 years and longer has been also associated with significantly increased risk of acoustic neuroma, a type of benign

brain tumor, on the primary side of cell phone use (International Agency for Research on Cancer (IARC) 2008; Lonn 2004; Schoemaker 2005). An extensive review of published studies of acoustic neuroma found that long-term cell phone users had a 60 percent greater risk of being diagnosed with the disease (Hardell 2009; Kundi 2009).

- A study from Israel reported an association between frequent and prolonged mobile phone use and parotid (salivary) gland tumors (Sadetzki 2008). Scientists analyzing data from Sweden and Denmark combined found that people who had used cell phones for at least 10 years ran an increased risk of benign parotid gland tumors (International Agency for Research on Cancer (IARC) 2008; Lonn 2006).

The National Research Council (NRC) has observed that "with the rapid advances in technologies and communications utilizing [radiation in the range of cell phone frequencies], children are increasingly exposed... at earlier ages (starting at age 6 or before)" (NRC 2008).

Research by France Telecom scientists showed that under standard conditions of use, twice as much cell phone radiation would penetrate a child's thinner, softer skull than an adult's (Wiart 2008). Children will be exposed to cell phone radiation for more years and therefore in greater total amounts than the current generation of adults (NRC 2008).

Children are likely to be more susceptible than adults to effects from cell phone radiation, since the brain of a child is still developing and its nervous tissues absorb a greater portion of incoming radiation compared to that of an adult (Conil 2008; de Salles 2006; Gandhi 1996; Kang 2002; Martinez-Burdalo 2004; Wang 2003; Wiart 2008).

Much more research is essential. However, in response to the information already available over the potential health risks of cell phone emissions, government agencies in Germany, Switzerland, Israel, United Kingdom, France, and Finland and the European Parliament have recommended actions to help consumers reduce exposures to cell phone radiation, especially for young children. Among warnings issued by government agencies are the following:

- United Kingdom Department of Health: "UK Chief Medical Officers strongly advise that where children and young people do use mobile phones, they should be encouraged to: use mobile phones for essential purposes only; keep all calls short - talking for long periods prolongs exposure and should be discouraged." (UK Department of Health 2005).

- Canada - City of Toronto Department of Public Health: "Today's children have started to use cell phones at a younger age, therefore their lifetime exposure to cell phone RFs will likely be greater. As a result, the chances that a child could develop harmful health effects from using a cell phone for a long time may be greater... Toronto Public Health is recommending that children, especially pre-adolescent children, use landlines whenever possible, keeping the use of cell phones for essential purposes only, limiting the length of cell phone calls and using headsets or hands-free options, whenever possible." (Toronto Public Health 2008a, b).
- Finland - Finnish Radiation and Nuclear Safety Authority: "It would be good to restrict children's use of mobile phones." "Precaution is recommended for children as all of the effects are not known... Parents are recommended to guide their children to use a handsfree that minimises the exposure of head significantly. When using a hands-free it is recommended to keep the mobile phone at least a few centimetres away from the body." (STUK (Finnish Radiation and Nuclear Safety Authority) 2009).

In contrast, the two U.S. federal agencies that regulate cell phones, the FDA and the Federal Communications Commission (FCC), have all but ignored evidence that long term cell phone use may be risky.

3. Federal Standards for Cell Phone Radiation Need to Be Modernized

The FCC set cell phone radiation standards 17 years ago, when few people used cell phones. These standards fail to provide an adequate margin of safety for cell phone radiation exposure and do not account for risks to children. The FCC standards closely follow the 1992 recommendations of the Institute of Electrical and Electronics Engineers (IEEE) (FCC 1997).

The FCC adopted IEEE's proposal to allow 20 times more radiation to the head than the average amount allowed for the whole body, even though the brain may well be one of the most sensitive parts of human body with respect to radiofrequency radiation and should have more protection.

EWG's conclusion: current U.S. cell phone radiation standards are outdated and may not be sufficiently protective. EWG urges the FDA and the FCC to upgrade its standards to take account of the newest scientific evidence and also increasing cell phone use by children.

4. What Consumers Can Do to Reduce Exposures to Cell Phone Radiation

EWG recommends a number of simple actions consumers can take to reduce exposures to cell phone radiation.

We recommend these simple precautionary measures until the science on cell phone risks is settled, and until the federal government modernizes current radiation limits to reflect the latest research.

- Use a low-radiation phone. Consumers can find radiation emissions for their current phone on EWG's database (www.ewg.org/cellphone - radiation), in their user's manual, or by contacting the manufacturer. EWG's database lists alternate, low-radiation phones, allowing people to consider purchasing a phone that emits the lowest radiation possible and still meets their needs.
- Use a headset or speakers. Headsets emit much less radiation than phones. Experts are split on whether wireless or wired is safer. Some wireless headsets emit continuous, low-level radiation, so EWG advises removing the headset from the ear between calls. Using a phone in speaker mode also reduces radiation to the head.
- Listen more, talk less. Cell phones emit radiation to transmit voice or text messages, but not to receive messages. Listening more and talking less reduces exposures.
- Hold phone away from the body. Holding the phone away from the torso when talking (while using the headset or speaker) reduces radiation exposures. EWG advises against holding the phone against the ear, in a pocket, or on the belt where soft body tissues absorb radiation.

- Choose texting over talking. Phones use less power (less radiation) to send text than voice. And unlike speaking with the phone at the ear, texting keeps radiation away from the head.
- Stay off the phone if the signal is poor. Fewer signal bars on the phone means that it emits more radiation to get the signal to the tower. EWG recommends that people make and take calls when the phone has a strong signal.
- Limit children's phone use. Young children's brains absorb twice the cell phone radiation as an adult's. EWG joins health agencies in at least six countries in recommending limits for children's phone use, such as for emergency situations only.
- Skip the "radiation shield." Radiation shields such as antenna caps or keypad covers reduce the connection quality and force the phone to transmit at a higher power with higher radiation.

5. Recommendations

The government should invest in additional research on the health effects of cell phone radiation, with special emphasis on children and teens. The government should require industry to make cell phone radiation level information available at the point of sale, so consumers can make informed decisions about the phones they buy.

Given the troubling questions raised by the research thus far, the cell phone industry should not wait for government action, but instead, offer consumers phones that operate with the least possible radiation, and should offer radiation information at the point of sale.

In the meanwhile, cell phone users can protect themselves and their families by buying low radiation phones. Cell phone users can also reduce radiation exposures by using their phone in speaker mode or with a headset.

In conclusion, EWG strongly believes that the government should support additional research into this important health question, and that the public has the right to know what levels of radiation they may be exposed to, what may be the potential risks, and what precautionary measures they can take to protect themselves and their families from any adverse health effects of cell phone radiation.

Thank you for your time. I welcome the opportunity to answer any questions you may have.

REFERENCES

Conil E, Hadjem A, Lacroux F, Wong MF, Wiart J. 2008. Variability analysis of SAR from 20 MHz to 2.4 GHz for different adult and child models using finite-difference time-domain. Phys Med Biol 53(6): 1511-25.

CTIA. 2009. Wireless Quick Facts. Year End Figures. Available: http://www.ctia.org/media/industry_info/index.cfm/AID/10323 [accessed September 2 2009]. de Salles AA, Bulla G, Rodriguez CE. 2006. Electromagnetic absorption in the head of adults and children due to mobile phone operation close to the head. Electromagn Biol Med 25(4): 349-60. FCC. 1997. Federal Communications Commission Office of Engineering & Technology. Evaluating Compliance with FCC Guidelines for Human Exposure to Radiofrequency Electromagnetic Fields. OET Bulletin 65 Edition 97-01 Available: http://www.fcc.gov/oet/rfsafety/ [accessed December 10 2008].

FDA. 2003. Cell Phone Facts. Consumer Information on Wireless Phones. Questions & Answers. Available: http://web.archive.org/web/ 20031205 121949/www.fda.gov/cellphones/qa.html [accessed June 10 2009].

Gandhi OP, Lazzi G, Furse CM. 1996. Electromagnetic absorption in the human head and neck for mobile telephones at 835 and 1900 MHz. IEEE Transactions on Microwave Theory and Techniques 44(10): 1884 - 97.

Hardell L, Carlberg M, Hansson Mild K. 2009. Epidemiological evidence for an association between use of wireless phones and tumor diseases. Pathophysiology 116(2-3): 113-22.

Hours M, Bernard M, Montestrucq L, Arslan M, Bergeret A, Deltour I, et al. 2007. [Cell Phones and Risk of brain and acoustic nerve tumours: the French INTERPHONE case-control study]. Rev Epidemiol Sante Publique 55(5): 321-32.

International Agency for Research on Cancer (IARC). 2008. INTERPHONE study results latest update - 8 October 2008. Available: http://www.iarc.fr/en/researchgroups/RAD/current-topics.php [accessed June 8, 2009].

ITU. 2008. International Telecommunication Union Press Release. Worldwide mobile cellular subscribers to reach 4 billion mark late 2008. Available: http://www.itu.int/newsroom/press_releases/2008/29.html [accessed March 19 2009].

Kang G, Gandhi OP. 2002. SARs for pocket-mounted mobile telephones at 835 and 1900 MHz. Phys Med Biol 47(23): 4301-13.

Kundi M. 2009. The Controversy about a Possible Relationship between Mobile Phone Use and Cancer. Environ Health Perspec 117(3): 316-24.

Lahkola A, Auvinen A, Raitanen J, Schoemaker MJ, Christensen HC, Feychting M, et al. 2007. Mobile phone use and risk of glioma in 5 North European countries. Int J Cancer 120(8): 1769-75.

Lonn S, Ahlbom A, Christensen HC, Johansen C, Schuz J, Edstrom S, et al. 2006. Mobile phone use and risk of parotid gland tumor. Am J Epidemiol 164(7): 637-43.

Lonn S, Ahlbom A, Hall P, Feychting M. 2004. Mobile phone use and the risk of acoustic neuroma. Epidemiology 15(6): 653-9.

Martinez-Burdalo M, Martin A, Anguiano M, Villar R. 2004. Comparison of FDTDcalculated specific absorption rate in adults and children when using a mobile phone at 900 and 1800 MHz. Phys Med Biol 49(2): 345-54.

NRC. 2008. National Research Council. Identification of Research Needs Relating to Potential Biological or Adverse Health Effects of Wireless Communication. Available: http://www.nap.edu/catalog.php?record_id=12036 [accessed December 10 2008].

Sadetzki S, Chetrit A, Jarus-Hakak A, Cardis E, Deutch Y, Duvdevani S, et al. 2008. Cellular phone use and risk of benign and malignant parotid gland tumors--a nationwide case-control study. Am J Epidemiol 167(4): 457-67.

Schoemaker MJ, Swerdlow AJ, Ahlbom A, Auvinen A, Blaasaas KG, Cardis E, et al. 2005. Mobile phone use and risk of acoustic neuroma: results of the Interphone case-control study in five North European countries. Br J Cancer 93(7): 842-8.

Schuz J, Bohler E, Berg G, Schlehofer B, Hettinger I, Schlaefer K, et al. 2006. Cellular phones, cordless phones, and the risks of glioma and meningioma (Interphone Study Group, Germany). Am J Epidemiol 163(6): 512-20.

STUK (Finnish Radiation and Nuclear Safety Authority). 2009. Statement of Finnish Radiation and Nuclear Safety Authority (STUK) concerning mobile phones and health on 7th January 2009. Available: http://www.stuk.fi/sateilytietoa/sateilyn_terveysvaikutukset/matkapuhelin_terveysvaikutu s/en_GB/stukin_matkapuhelinkannanotto/ [accessed March 20 2009].

Toronto Public Health. 2008a. Cell Phone Use by Children and Youth. Board of Health report (May 2008). Available: http://www.toronto.ca/health/hphe/radiation/radiofrequency.htm [accessed March 29 2009].

Toronto Public Health. 2008b. Fact Sheet 9 Children and Safe Cell Phone Use (July 2008). Available: http://www.toronto.ca/health/hphe/radiation/radiofrequency.htm [accessed March 29 2009].

UK Department of Health. 2005. Mobile Phones and Health. Available: http://www.dh.gov.uk/en/Publicationsandstatistics/Publications/PublicationsPolicyAndGu idance/DH_4006938 [accessed March 29 2009].

Wang J, Fujiwara O. 2003. Comparison and Evaluation of Electromagnetic Absorption Characteristics in Realistic Human Head Models of Adult and Children for 900-MHz Mobile Telephones IEEE Transactions on Microwave Theory and Techniques 51(3): 966970.

Wiart J, Hadjem A, Wong MF, Bloch I. 2008. Analysis of RF exposure in the head tissues of children and adults. Phys Med Biol 53(13): 3681995.

INDEX

#

A

B

C

D

E

F

G

H

I

S

T